# Forest Management Auditing

Forest management auditing is expanding from its traditional focus on forest management, stewardship and Chain of Custody certification to more innovative topics such as ecosystem services, forest carbon credits, Non-Wood Forest Products, wood energy and Fair Trade certification. Forest managers, auditors and project developers require a new range of skills, capacities and expertise to monitor these issues. This book outlines the market-based tools that are required by such professionals to ensure corporate social responsibility in the forestry sector.

It shows how a mutual learning process between established forest certification processes and innovative markets is needed. It addresses key topics such as High Conservation Value (HCV) approaches, the role of independent certification versus due diligence process and the engagement of smallholders and SMEs. Beginning with a market and policy analysis, the book fosters a deeper understanding of standards, methodologies and auditing techniques. Numerous case studies are included from a wide range of contexts, including both temperate and tropical forests, developed and developing countries. Overall, the book analyses all the steps towards forest management and forest products and services certification.

**Lucio Brotto** is Co-Founder and Director of the Sustainable Investments Unit at ETIFOR, a spin-off company of the University of Padova, Italy. He is active in developing and auditing forest-based projects and has experience in Europe, Costa Rica, Honduras, Haiti, Colombia, Peru, Brazil, Argentina, DR Congo, Uganda, Angola, Kenya, Morocco, Tunisia, Cambodia and Vietnam. Typically Lucio is involved with sustainable planning and management of forests according to voluntary certification standards such as the FSC and carbon standards.

**Davide Pettenella** is Professor of Forest Policy and Economics at the Department of Land, Environment, Agriculture and Forestry of the University of Padova, Italy. He has worked with numerous organizations including the World Bank, FAO, European Commission, Ministry of Foreign Affairs and ENCC in developing countries and economies in transition.

Earthscan Studies in Natural Resource Management

**Global Resource Scarcity**
Catalyst for conflict or cooperation?
*Edited by Marcelle C. Dawson, Christopher Rosin and Navé Wald*

**Socio-Ecological Resilience to Climate Change in a Fragile Ecosystem**
The Case of the Lake Chilwa Basin, Malawi
*Edited by Sosten Chiotha, Tembo Chanyenga, Joseph Nagoli, Patrick Likongwe and Daniel Jamu*

**Southern African Landscapes and Environmental Change**
*Edited by Peter Holmes and John Boardman*

**Large Carnivore Conservation and Management**
Human Dimensions
*Edited by Tasos Hovardas*

**Forest Management Auditing**
Certification of forest products and services
*Edited by Lucio Brotto and Davide Pettenella*

**Agricultural Land Use and Natural Gas Extraction Conflicts**
A Global Socio-Legal Perspective
*Madeline Taylor and Tina Hunter*

**Community Based Natural Resource Management**
From economic principles to practical governance
*Brian Child*

For more information on books in the Earthscan Studies in Natural Resource Management series, please visit the series page on the Routledge website: http://www.routledge.com/books/series/ECNRM/

# Forest Management Auditing

## Certification of Forest Products and Services

Edited by Lucio Brotto and Davide Pettenella

LONDON AND NEW YORK

from Routledge

First published 2018 by Routledge

2 Park Square, Milton Park, Abingdon, Oxfordshire OX14 4RN
52 Vanderbilt Avenue, New York, NY 10017

*Routledge is an imprint of the Taylor & Francis Group, an informa business*

First issued in paperback 2020



*British Library Cataloguing-in-Publication Data*
A catalogue record for this book is available from the British Library

*Library of Congress Cataloging-in-Publication Data*
Names: Brotto, Lucio, editor. | Pettenella, Davide, editor.
Title: Forest management auditing : certification of forest products and services / edited by Lucio Brotto and Davide Pettenella.
Description: Milton Park, Abingdon, Oxon ; New York, NY : Routledge, 2018. | Includes bibliographical references and index.
Identifiers: LCCN 2018018609 | ISBN 9781138816671 (hbk) | ISBN 9781315745985 (ebk)
Subjects: LCSH: Forest management—Auditing. | Forest management—Standards. | Forest products—Certification.
Classification: LCC SD431 .F668 2018 | DDC 634.9/20681—dc23
LC record available at https://lccn.loc.gov/2018018609

ISBN: 978-1-138-81667-1 (hbk)
ISBN: 978-0-367-60587-2 (pbk)

Typeset in Sabon
by Apex CoVantage, LLC

# Contents

# Figures

# Tables

# Boxes

# Acronyms and abbreviations

| | |
|---|---|
| A/R | Afforestation and reforestation |
| ACR | American Carbon Registry |
| AFOLU | Agriculture, Forestry and Other Land Use |
| ASI | Accreditation Service International |
| BBOP | Business and Biodiversity Offset Program |
| CAR | Climate Action Reserve |
| CB | Certification body |
| CCB | Climate, Community and Biodiversity Standard |
| CDM | Clean Development Mechanism |
| CERs | Certified Emission Reductions |
| CGF | Consumer Goods Forum |
| CO2 | Carbon dioxide |
| CoC | Chain of Custody |
| CSR | Corporate social responsibility |
| DDS | Due Diligence System |
| EIA | Environmental Impact Assessment |
| EMS | Environmental Management Systems |
| ETS | Emissions Trading System |
| EU | European Union |
| EUTR | European Union Timber Regulation |
| FAO | Food and Agriculture Organization of the United Nations |
| FLEGT | Forest Law Enforcement, Governance and Trade |
| FLO | Fairtrade Labelling Organization |
| FM | Forest Management |
| FME | Forest Management Entity |
| FMO | Forest Management Operation |
| FMU | Forest Management Unit |
| FSC | Forest Stewardship Council |
| GACP | Good Agricultural and Collection Practices |
| GRI | Global Reporting Initiative |
| GHG | Greenhouse gases |
| HCV | High Conservation Values |

| | |
|---|---|
| ICROA | International Carbon Reduction and Offset Alliance |
| IFC | International Finance Corporation |
| IFM | Improved Forest Management |
| IFOAM | International Federation of Organic Agriculture Movements |
| IRCA | International Register of Certificated Auditors |
| ISO | International Organization for Standardization |
| ISSC-MAP | International Standard for Sustainable Wild Collection of Medicinal and Aromatic Plants |
| JI | Joint Implementation |
| LCA | Life Cycle Assessment |
| LULUCF | Land Use, Land Use Change and Forestry |
| MOs | Monitoring organizations |
| NFSS | National Forest Stewardship Standards |
| NGO | Non-governmental organization |
| NWFPs | Non-Wood Forest Products |
| NYDF | New York Declaration on Forests |
| P&C | Principles and criteria |
| PDD | Project Design Document |
| PDO | Protected Designation of Origin |
| PEFC | Program for the Endorsement of Forest Certification |
| PES | Payment for Ecosystem Services |
| PGI | Protected Geographical Indication |
| PIN | Project Idea Note |
| PWS | Payments for Watershed Services |
| REDD | Reducing Emissions from Deforestation and Forest Degradation |
| RSPO | Roundtable on Sustainable Palm Oil |
| SBP | Sustainable Biomass Program |
| SFI | Sustainable Forestry Initiative |
| SFM | Sustainable forest management |
| SLIMF | Small and Low Intensive Managed Forests |
| SMEs | Small and medium enterprises |
| TSG | Traditional Speciality Guaranteed |
| UNFCCC | United Nations Framework Convention on Climate Change |
| VCS | Verified Carbon Standard |
| VERs | Verified Emission Reductions |
| VPAs | Voluntary Partnership Agreements |
| VVB | Validation/verification body |
| WBCSD | World Business Council for Sustainable Development |
| WFTO | World Fair Trade Organization |
| WHO | World Health Organization |
| WQT | Water Quality Trading |
| WTO | World Trade Organization |
| WWF | World Wildlife Fund |

# Contributors

**Nicola Andrighetto**: 2017–present Responsible Management Unit Director at ETIFOR; 2011–present Consultant on Wood Energy and Timber Legality at University of Padova.

**Lucio Brotto**: 2011–present Co-founder and Sustainable Investments Unit Director at ETIFOR; 2009–present Consultant on Forest Carbon Market at University of Padova; 2012–2015 PhD Fellow at University of Padova & University of Dresden; 2008 Assistant Project Leader at UN Economic Commission of Europe/FAO.

**Ariadna Chavarria**: 2017–present Applied Scientist on Wastewater Treatment and Regulations at E3 Solutions Tampa, USA; 2012–2016 PhD Fellow at University of Padova.

**Giulia Corradini**: 2016–present External Researcher at University of Padova in Forest Policy and Economics.

**Oliver David Miles Cupit**: 2014–present Consultant at NEPCon.

**Diego Florian**: 2011–present FSC Italy Director; 2009–2012 External Researcher at University of Padova; 2008 Fellow at FAO-Mozambique on National Food Security Program.

**Mateo Cariño Fraisse**: 2004–present Forest and Climate Services Coordinator and Lead Auditor at NEPCon; 2002–2004 Forest Consultant.

**Alessandro Leonardi**: 2011–present Co-founder and Managing Director at ETIFOR; 2016–2017 Consultant at UNECE/FAO Forestry and Timber Committee; 2012–2014 PhD Fellow at University of Padova; 2010–2012 EU Projects Coordinator at Fundacion COPADE.

**Aynur Mammadova**: 2016–present PhD Fellow at the University of Padova.

**Mauro Masiero**: 2017–present Assistant Professor University of Padova; 2014–2016 Post-Doc at University of Padova; 2011–2014 PhD Candidate at University of Padova; 2003–2011 External Researcher at University of

Padova; 2005–2011 FSC Italy Secretariat; 2002–2005 QualiTree Forest Certification Consultant; 2003–2005 Lead FSC Auditor Certiquality.

**Davide Pettenella**: 2014–present Coordinator of the PhD school LERH; 1990–present Professor of Forest Policy and Economics at University of Padova; 1980–1990 Researcher at Centro di Sperimentazione Agricola e Forestale (SAF).

**Alex Pra**: 2016–present Sustainable Investment Officer at ETIFOR; 2015–present PhD Fellow at University of Padova.

**Laura Secco**: 2014–present Associate Professor at University of Padova; 2001–2005 FSC Italy National Secretary and 1998–2001 National Contact Person; 2005–2013 Researcher and 1996–2005 Research Assistant at University of Padova and EFI.

**Ondřej Tarabus**: 2013–present Biomass Program Manager at NEPCon.

**Sabrina Tomasini**: 2014–present PhD Fellow at University of Copenhagen and University of Padova; 2014–2016 co-editor of Policy Brief Series at the Copenhagen Centre for Development Research (CCDR); 2013 Assistant Project Coordinator at WHO Traditional Medicine Unit; 2012 Research Assistant at World Agroforestry Centre (ICRAF).

**Ilaria Dalla Vecchia**: 2015–present Technical Expert at FSC Italy.

**Enrico Vidale**: 2011–present External Consultant at University of Padova on Wild Forest Products.

Chapter 1

# Cutting edge

## Sustainability in the forestry sector

*Lucio Brotto, Davide Pettenella and Alex Pra*

The attention towards sustainability in the forestry sector has been growing at unprecedented levels in the last decades, accompanied by the emergence of corporate social responsibility (CSR) as a key tool to address sustainability challenges. There are several definitions of CSR in literature. In the context of this book, we make reference to the definition given by the European Commission (2011), which defines CSR as "a concept whereby companies integrate social and environmental concerns in their business operations and in their interactions with their stakeholders on a voluntary basis" (p. 6). The concept of CSR first appeared in scientific studies in the 1980s (Carroll, 1999). Since its initial focus on social aspects, nowadays CSR is increasingly aligned with the concept of sustainability, encompassing a broader range of business aspects such as environment, governance and economics (Vidal and Kozak, 2008). Different approaches to the relationship between business and society have generated different ways of looking at CSR (Han, 2010):

- Instrumental theories: CSR is considered only as a tool to achieve economic goals, hence long-term profit maximization (Friedman, 1970);
- Political theories: business is powerful and can impact society; the more social power there is in a business, the more social responsibilities there are (Davis, 1960);
- Integrative theories: business depends on society and a company should listen and integrate social demand (Preston and Post, 1975);
- Ethical theories: ethical standards are facilitating good business-society relationships; the company is committed not only to shareholders but also to stakeholders (Freeman, 1984).

In the forestry sector, CSR instruments emerged in the early 1990s (Cashore, 2002), to face the growing public interest and concern about environmental and social problems related to forests, in particular, those problems linked to the core issues of deforestation and degradation processes and forest illegality. The awareness towards these issues has been growing substantially since the 1990s, together with the number of companies and organizations engaging with CSR tools to increase transparency, minimize reputational risks and reduce costs connected to lawsuits and boycott campaigns; and also to gain market competitiveness (e.g. avoid loss of market share, enter new markets and obtain price premium), improve reputation and legitimacy and integrate stakeholders' interest through win-win synergic value creation activities (Kurucz et al., 2008; Jenkins and Smith, 1999; Vidal and Kozak, 2008).

From a more global perspective, the self-regulation process of the private sector has been driven by a number of factors: (i) the growing difficulties of governments in regulating and monitoring transnational corporations and the financial market; (ii) the failure of policy instruments (command and control instruments) in promoting the sustainable management of natural resources; (iii) the "rolling back the frontiers of the State" with a transfer of environmental and social decisions from State level to corporate sphere (Heal, 2008); (iv) an increased role of civil society in decision-making, shifting from a "government" to a "governance" level; and (v) the internationalization of companies with the shifting of operations in less developed countries characterized by poor law enforcement and fragile social situations (Heal, 2008; Voegtlin et al., 2011; Zhang et al., 2014).

Forest certification, i.e., the third-party control of the responsible management of forest resources and the tracking of forest products from the forest to the final consumers, has emerged as one of the leading voluntary CSR tools in the forestry sector. The book by Viana et al., *Certification of Forest Products: Issues and Perspectives*, published in 1996, begins as follows: "Certification of forest products is a new and exciting phenomenon. Its future is still uncertain, but it could have a significant impact on forest practices around the world in coming years". After more than 20 years, we can say that the certification of forests and forest products is a consolidated tool of corporate social responsibility (CSR) incorporating almost one-sixth of the world's forest cover (UNECE, 2016), tens of thousands of processing companies and millions of consumers around the world. Moreover, even if the Ecolabel Index database[1] is recording 16 forest-related ecolabels used in the forestry sector, two standards and their certification procedures are dominating the market and are by far more visible than others. CSR tools in the forestry sector are rapidly evolving far beyond the use of standards related to responsible forest management and related independent systems of control. First of all, there has been an enlargement of the area of concern of the standards, with new aspects considered and with new certification systems developed: environmental services, Non-Wood Forest Products including, also, edible and medicinal products and woody biomass to be used for energy production or in the building sector. There has also been an enlargement of the sources of information about the performance of the companies operating in the sector – data and information are collected not only with external desk and field audits, but also through reporting by stakeholders with mobile technology, through remote sensing based on fine resolution images, through sensors placed inside the forest. The origin of the forest products can be investigated with wood and fibre testing (both traditional wood anatomy and DNA/isotope future capacity) and can be tracked with blockchain technology (Dudder and Ross, 2017).

In the last years, certification has been progressively associated to the broadening scope of CSR strategies to social, environmental, economic and governance aspects (Toppinen and Zhang, 2010) with the disclosure and reporting of sustainability performances associated to new social demands by consumers, public authorities and investors (see, for example, the development of impact investments associated to the need of generating positive and measurable social and environmental impacts alongside financial returns). This, in the framework of CSR, is a general trend in all the economic sectors, but forestry represents an interesting – and, from many perspectives, singular – case for two main reasons. Firstly, it represents an emblematic case where the private sector not only anticipated public institution initiatives (it was not until the beginning of the 2000s that the need to address forestry sustainability challenges was reflected in the international forestry policy agenda and regulatory initiatives were implemented), but in many cases, voluntary CSR tools functioned as a catalyser of approaches and methodologies that have been then adapted or have inspired public institutions initiatives. Secondly, few other large primary sectors have experienced such a high number of initiatives to address legality and sustainability issues, creating an unbalanced competitive condition among wood products and alternative raw materials in which CSR standards and requirements are not developed and used (plastic, metals, fossil fuels, etc.).

Looking at this complex and dynamic context, the book aims to outline the innovative market-based tools that are shaping CSR in the forestry sector. Starting from a market and policy analysis, the book tries to capture the quickly evolving world of standards, methodologies and auditing techniques applied in the forestry sector. The book analyses all the steps towards forest management and forest products and services certification, supported by case studies and tools as examples, providing auditors and project developers with new multiple capacities and knowledge.

The book is organized in three main parts. The first part introduces the definitions and the overall framework of CSR in the forestry sector. The second part presents the main standards and systems for the certification of forest products and services. The third part complements the previous by presenting the auditing techniques related to the main standards. The structure of the book is presented in Figure 1.1.

Chapters 2 to 5 compose the first part of the book, describing the driving forces behind the uptake of CSR in the forestry sector and the evolution of

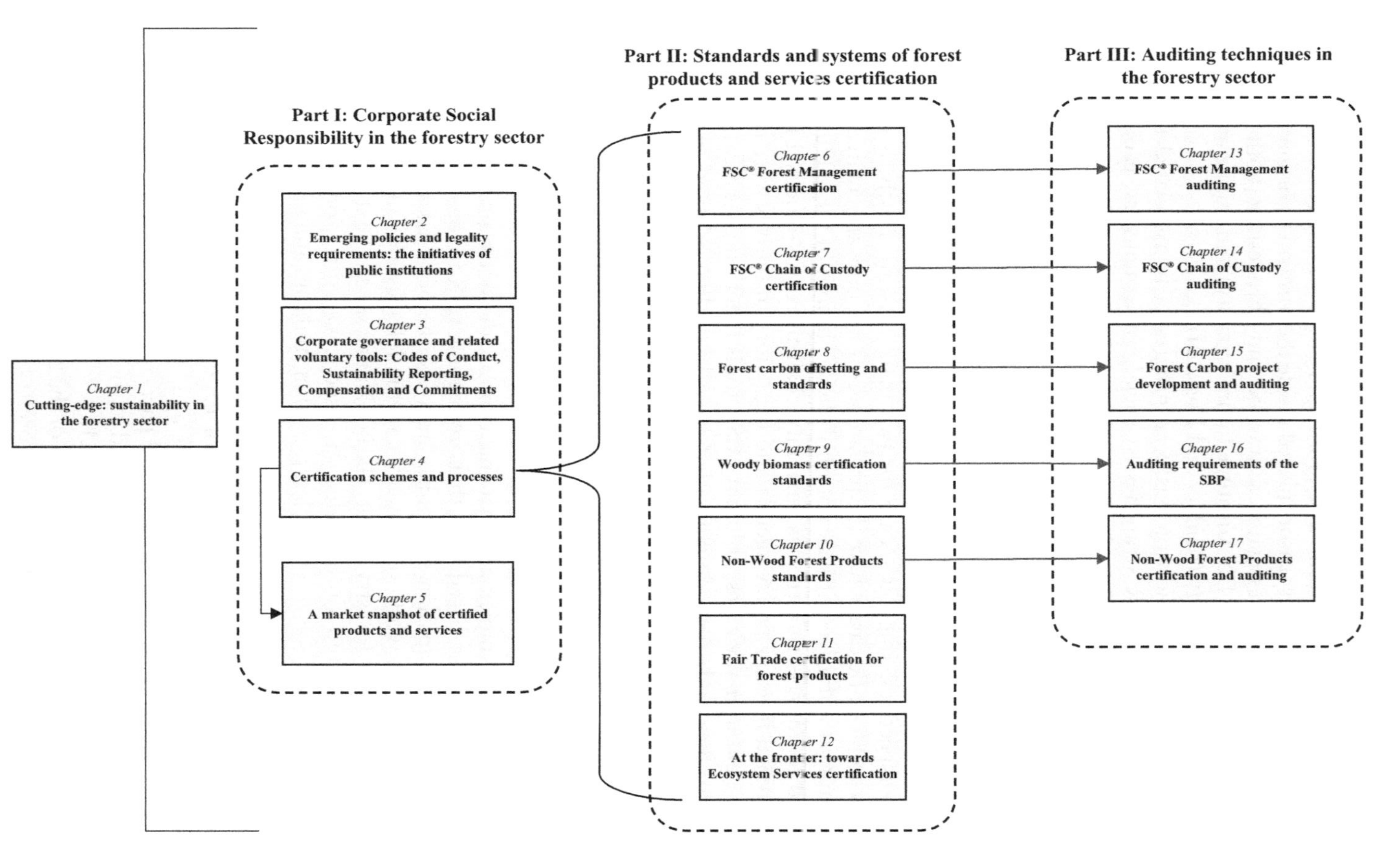


*Figure 1.1* Book structure

Source: Own elaboration

the regulatory initiatives of public institutions on the one hand and voluntary CSR tools on the other.

Chapter 2 presents the main policy developments since the beginning of 2000s to tackle illegality in the forestry sector and the main initiatives of public institutions aiming to exclude illegal timber from domestic/regional markets. Particular attention is put on the European Union (EU) Forest Law Enforcement, Governance and Trade (FLEGT) Action Plan and the EU Timber Regulation (EUTR) in Europe, the Lacey Act in the United States and the Australian Illegal Logging Prohibition Bill in Australia, discussing the failures and successes of these systems. While Chapter 2 focuses on the public institutions initiatives with regulatory character, Chapters 3 and 4 focus on the voluntary tools that have emerged from the private sphere. The self-regulation process of the private sector, in fact, manifested in the spread of a wide range of CSR tools, from so-called soft tools such as corporate norms, codes of conduct and commitments, to certification schemes and processes. Chapter 3 provides an overview of "soft tools", distinguishing among: (i) codes of conduct (or codes of ethics), the early generation of corporate self-regulation tools; (ii) corporate reporting, the new generation of corporate self-regulation tools focusing more on external stakeholders, providing also examples of the main international guidelines on reporting (e.g., OECD Guidelines for Multinational Enterprises, United Nations Global Compact); (iii) compensation and offsetting measures; and finally (iv) commitments, an emerging tools where the attention is completely directed to external stakeholder and upstream suppliers. Special attention is provided to the latter tool, describing types and characteristics, implementation strategies and shortcomings and challenges. Chapter 4 presents certification schemes and processes applicable to the forestry sector, which represent the next-level corporate self-regulation tools, and is the core aspect of the book. Forest certification schemes are described starting from the traditional forest management and Chain of Custody (CoC) initiatives – Forest Stewardship Council (FSC®) and the Program for Endorsement of Forest Certification schemes (PEFC) – moving to the other management schemes not specific to the forestry sector but frequently used by forest organizations (ISO 9001 and 14001, BS OHSAS 18001, etc.) and those related to innovative sectors such as forest carbon offsetting, wood energy, Non-Wood Forest Products (NWFPs), Fair Trade and the newer frontier of ecosystem services certification. The chapter introduces the general concepts, actors and approaches related to certification schemes and processes as well as providing an overview of the contents of the main certification schemes that can be used in the forestry sector. The first part of the book ends with Chapter 5, which provides a snapshot of the markets for certified products and services, focusing on the markets of FSC-certified products, including analysis of the reasons for certification by businesses, public administration and non-governmental organizations (NGOs) – carbon finance, fuelwoods markets, NWFP markets and markets for other services such as watershed and biodiversity.

The second part of the book includes Chapters 6 to 12 and analyses in detail the standards and the certification infrastructures for the certification of forest products and services:

- FSC Forest Management Certification (Chapter 6);
- FSC CoC certification (Chapter 7);
- forest carbon standards (Chapter 8);
- woody biomass for energy standards (Chapter 9);
- NWFP standards (Chapter 10);
- Fair Trade certification of forest products (Chapter 11);
- standards and certification-integrations for ecosystem services (Chapter 12).

Standards are presented and analysed based on the structure of their principles, criteria and indicators with additional references to standard documentation such as such policy documents, guidance notes, etc. The certification infrastructure of each process is analysed considering the standard-setting bodies, certification bodies and accreditation services.

Chapter 6 presents the FSC Forest Management Certification system, presenting the overall framework of the latest standard and the main requirements for addressing socio-economic impacts, environmental impacts, stakeholder consultation and management of High Conservation Value (HCV) forests. In addition, the special tools and approaches for smallholders, such as the Group Certification and Small and Low Intensity Managed Forests (SLIMF), are described. Following this, Chapter 7 presents CoC certification according to FSC, introducing the key concept of CoC in relation to forest supply chains and presenting the requirements according to FSC's latest CoC standard. In addition, it provides an overview of systems for controlling FSC claims, the FSC Controlled Wood standard, multisite/group certification and FSC labelling requirements. Chapter 8 provides an overview of standards applicable to forest-based carbon projects. It introduces the concept of carbon offsetting and the markets for trading forest carbon offsets, both compliance and voluntary. The future developments connected with the entering into force of the recent Paris Agreement are also discussed. The chapter describes the types of forest-based projects, i.e., afforestation and reforestation projects, Improved Forest Management and Reducing Emissions from Deforestation and Forest Degradation (REDD) projects, including the key elements that standards need to address, i.e., additionality, permanence, leakage, double counting, etc. The chapter provides a comparison of the main standards used for forest carbon projects, such as the Verified Carbon Standard (VCS), the Climate, Community and Biodiversity (CCB) standard, the Gold Standard, Plan Vivo system, the American

Carbon Registry as well as UK Woodland Carbon Code. Chapter 9 focuses on woody biomass for energy standards, a growing sector in terms of market as well as public concern, with a growing number of sustainability initiatives and standards. This chapter presents the sustainability requirements for some key countries and existing certification schemes. Special attention is put on the Sustainable Biomass Partnership (SBP) certification. Finally, the ENPlus® scheme, as an example of quality standard, is also presented. Apart from wood and wood-based products, certification can also apply to collection, production and processing of NWFPs such as greeneries, mushrooms, berries, nuts, resins, essential oils, litter, medicinal plants, etc. Chapter 10 presents the standards and certification schemes most relevant for NWFP certification, from forest management and CoC standards, which covers most of the NWFPs, to *ad hoc* standards. Chapter 11 presents the recent initiatives for the Fair Trade certification of forest products, aimed specifically at providing certification solutions for smallholders and community forestry in the context of less developed countries. The chapter presents the initiatives directly developed by forest certification schemes, e.g., the Small and Community Label Option (SCLO) initiative under FSC, and those launched in cooperation with other organizations, i.e., the pilot initiatives between forest certification and Fair Trade schemes, namely the FSC–Fairtrade dual certification pilot project.

Up to this point, the consolidated standards and certification systems in the forestry sector have been presented; Chapter 12 provides a close look at the new frontier of ecosystem services certification. The chapter focuses on the FSC certification of ecosystem services, based on the Forest Certification for Ecosystem Services (ForCES) pilot project and other certification initiatives integrating ecosystem services other than carbon, such as biodiversity and water-related services.

Chapters 13 to 17 compose the third and final part of the book, highlighting auditing requirements, auditing techniques, project cycles and certification procedures for five selected processes of forest products and services certification:

- FSC forest management auditing (Chapter 13);
- FSC Chain of Custody auditing (Chapter 14);
- forest carbon project development and auditing (Chapter 15);
- Sustainable Biomass Partnership auditing (Chapter 16);
- NWFPs certification and auditing (Chapter 17).

Chapter 13 complements Chapter 6 and provides a detailed and practical look into the processes of FSC forest management auditing, presenting auditing requirements and processes, from the initial contact to the reporting phase. Chapter 14 complements Chapter 7, presenting auditing standards and techniques for the assessment of CoC according to FSC distinguishing them into CoC evaluation at forest management level and CoC evaluation as a separate assessment. Chapter 15 complements Chapter 8 and has a more general approach compared to the two previous chapters. This chapter provides a look into forest carbon project cycle, with a specific emphasis on auditing stages and procedures, making reference to the main forest carbon standards, from the Clean Development Mechanism (CDM) to VCS, CCB, Gold Standard and Plan Vivo. It includes also the main requirements for third-party auditor's accreditation and approval under these standards. Chapter 16 provides a detailed look into the processes of SBP auditing requirements. SBP is becoming the leading certification system used for solid biomass sustainability certification for industrial and large-scale energy production systems and is required by the main European utilities to prove compliance with the national legislation to the regulators. Finally, while Chapter 10 provides a description of the standards applicable to NWFPs, Chapter 17 provides an overview of the scopes of certification in different NWFP standards and clarifies two features that are essential for establishing meaningful auditing procedures: the ecological position and the habitat. This chapter ends the book by providing potential auditing indicators for the different NWFPs.

There is widespread social demand for transparent sustainability claims, concrete commitments and real responsible practices by companies and other organizations involved in the supply of forest products and services. We hope, with this book, to be able to support the development of these initiatives and to respond to the demands of trust and credibility expressed by a growing number of citizen-consumers much worried about the state of the world's forest resources.

## Note

1 Ecolabel Index. http://www.ecolabelindex.com

## References

Carroll, A.B. (1999). Evolution of a Definitional Construct. *Business and Society*, 38(3), 268–295.

Cashore, B. (2002). Legitimacy and the Privatization of Environmental Governance: How Non-State Market-Driven (NSMD) Governance Systems Gain Rule-Making Authority. *Governance*, 15(4), 503–529. doi:10.1111/1468-0491.00199

Davis, K. (1960). Can Business Afford to Ignore Social Responsibilities? *California Management Review*, 2(3), 70–76.

Dudder, B., Ross, O. (2017). Timber Tracking Reducing Complexity of Due Diligence by Using Blockchain Technology. In: B. Johansson (ed.), Joint Proceedings of the BIR 2017 pre-BIR Forum. Copenhagen, Denmark, August 28–30, 2017. http://ceur-ws.org/Vol-1898/paper15.pdf

European Commission (2011). *Corporate Social Responsibility: National Public Policies in the EU*. Brussels: European Commission.

Freeman, R.E. (1984). *Strategic Management: A Stakeholder Approach*. Boston: Pitman.
Friedman, M. (1970). The Social Responsibility of Business Is to Increase Its Profits. *The New York Times Magazine*. September 13.
Han, X. (2010). *Corporate Social Responsibility and Its Implementation: A Study of Companies in the Global Forest Sector*. Corvallis: Oregon State University.
Heal, G.M. (2008). *When Principles Pay: Corporate Social Responsibility and the Bottom Line* (p. 271). Columbia University Press. Retrieved from http://books.google.com/books?hl=en&lr=&id=_MPqLqSp79EC&pgis=1
Jenkins, M., Smith, E. (1999). *The Business of Sustainable Forestry: Strategies for an Industry in Transition* (p. 350). Island Press. Retrieved from http://books.google.com/books?id=N4omnNnzKooC&pgis=1
Kurucz, E., Colbert, B., Wheeler, D. (2008). The Business Case for Corporate Social Responsibility. In A. Crane, A. McWilliams, D. Matten, J. Moon, D. Siegel (Eds.), *The Oxford Handbook of Corporate Social Responsibility* (pp. 83–112). Oxford: Oxford University Press. ISBN: 978-0-19-921159-3.
Preston, L.E., Post, J.E. (1975). *Private Management and Public Policy: The Principle of Public Responsibility* (p. 175). Retrieved from http://books.google.it/books/about/Private_management_and_public_policy.html?id=B7QTAQAAMAAJ&pgis=1
Toppinen, A., Zhang, Y. (2010). Changes in Global Markets for Forest Products and Timberlands. In *Forest and Society: Responding to Global Drivers of Change* (pp. 135–156). Vienna: IUFRO World Series Vol. 25to.
UNECE/FAO (2016). *Forest Products Annual Market Review, 2015–2016*. United Nations Economic Commission for Europe, Geneva (p. 141). Retrieved from www.unece.org/fileadmin/DAM/timber/publications/fpamr2016.pdf
Vidal, N., Kozak, R. (2008). The Recent Evolution of Corporate Responsibility Practices in the Forestry Sector. *International Forestry Review*, 10(1), 1–13. doi:10.1505/ifor.10.1.1
Voegtlin, C., Patzer, M., Scherer, A.G. (2011). Responsible Leadership in Global Business: A New Approach to Leadership and Its Multi-Level Outcomes. *Journal of Business Ethics*, 105(1), 1–16. doi:10.1007/s10551-011-0952-4
Zhang, Y., Toppinen, A., Uusivuori, J. (2014). Internationalization of the Forest Products Industry: A Synthesis of Literature and Implications for Future Research. *Forest Policy and Economics*, 38, 8–16. doi:10.1016/j.forpol.2013.06.017

# Part I

# Corporate social responsibility in the forestry sector

# Chapter 2

# Emerging policies and legality requirements

## The initiatives of public institutions

*Mauro Masiero, Oliver Cupit and Davide Pettenella*

Whilst interest in deforestation and forest illegality dates back to the 1970s and gained momentum throughout the 1990s, it was not until the early part of the 21st century that the need to address illegality was reflected in the international forestry policy agenda and national initiatives to tackle forest crime were implemented by key states. This chapter provides an overview of the main policy developments designed to tackle illegality in the forestry sector in recent times by public institutions. Special attention will be paid to the European Union (EU) Forest Law Enforcement, Governance and Trade (FLEGT) Action Plan, with particular reference to Voluntary Partnership Agreements (VPAs) and to national and regional initiatives aimed at excluding illegal timber from domestic markets (such as the Lacey Act, EU Timber Regulation and the Australian Illegal Logging Prohibition Bill) (paragraph 2.3). Finally, some considerations regarding the pros and cons of such policies will be posited, including consideration of nascent impacts on timber markets and forest certification schemes (paragraph 2.4).

### 2.1 A historical overview of forest policy and legislation

Nascent initiatives aimed at addressing deforestation and forest illegality in the 1970s were led by the international scientific community and prominent environmental lobby groups. The efforts of these groups were concentrated on tackling the illegal trade in endangered species, predominantly in biodiverse habitats such as tropical forests. The Convention on International Trade in Endangered Species of Wild Fauna and Flora (CITES) entered into force in 1975, enshrining these aims in international law. In the same year, the World Wildlife Fund (WWF) launched the first tropical forest campaign, aimed at creating forest reserves in Africa, Asia and Latin America. Around this time the United Nations (UN) Food and Agriculture Organization (FAO) also published its first Tropical Rainforest Report. Whilst the objective of these initiatives was laudable, delivery of significant results was stymied by lack of institutional collaboration, shared strategy or a supportive policy framework. Nevertheless, together with boycott campaigns in the 1980s, they contributed to bringing deforestation and illegal logging issues to the public eye, preparing the ground for future action.

In 1992, the failure of the UN Earth Summit to deliver a global forest convention was broadly seen by experts as a catalyst for the development of various bilateral activities in forestry. It also acted as a key driver to private sector initiatives, such as forest certification (FERN, 2013). Despite its failures, this summit laid the groundwork for future innovation in policy, directly and indirectly influencing wider environmental initiatives, such as the Convention on Biological Diversity and the UN Convention to Combat Desertification. Illegal logging was only on the periphery of the international policy agenda in the early 1990s, with several countries denying that a problem even existed. However, by the latter part of the decade intergovernmental bodies had started to tackle illegal logging issues head-on, and new, more effective cooperation between the public sector, private sector and civil society began to develop. This increasingly collaborative and interdisciplinary approach enhanced the debate on illegal logging, focusing not only on narrow drivers of deforestation, but on underlying governance and law enforcement issues. In 1997, members of the G8 agreed at a summit in Denver, USA, to launch an Action Plan on Forests that included illegal logging among its five key issues. This plan, finalized in 2002, aimed to explore the scale and impacts of illegal logging in order to inform future policy on effective counter-measures (G8, 2002). As a result, G8 members committed to keeping forest-related issues high on the political agenda and to combat illegal logging (FERN, 2013). The G8 Action Plan was also implemented cooperatively with the World Bank's Forest Governance Program. Working through an operational partnership on forest law enforcement in South-East Asia, these two initiatives resulted in the 2001 Bali Declaration and the inception of the World Bank's Forest Law Enforcement and Governance (FLEG) regional initiatives. FLEG initiatives were ultimately extended to Eastern Asia and the Pacific (launched in 2001), Africa (launched in 2003), Europe and Northern Asia (launched in 2005) and Latin America and the Caribbean (launched in 2006). Off the back of these initiatives the EU, as one of the major wood consumers and importers worldwide, published its FLEGT Action Plan in 2003 (see 2.2 below). This was further refined, by means of specific regulations, in 2005 and 2008. The FLEGT Action Plan includes several measures to address forest crime, including the EUTR, which was approved by the EU Parliament in 2010. Together with the 2008 amendment to the US Lacey Act, the development of the EUTR signalled a new phase of demand-side measures to address trade in illegal timber. In 2012, the Australian Parliament followed suit, approving the Australian Illegal Logging Prohibition Act. Other governments are now in the process of developing similar laws or adapting existing legislation to address trade in illegal timber. Although "sustainable" forest management remains high on the policy agenda, timber "legality" is the new hot topic, representing, as it does, a necessary prerequisite for responsible management of the world's forests.

## 2.2 The Forest Law Enforcement Governance and Trade (FLEGT) action plan

The EU published the FLEGT Action Plan in 2003. The initiative is aimed at preventing the import of illegal timber into the EU, improving the supply of legal timber to the EU market and increasing demand for timber from responsibly managed forests. As such, the FLEGT Action Plan is to be considered amongst a cadre of efforts implemented by the EU, aimed at promoting sustainable forest management (EC, 2003). It is composed of seven key components (Box 2.1).

**Box 2.1 Key areas of the EU FLEGT Action Plan**

The EU FLEGT Action Plan includes the following key action areas (EC, 2003):

1. supporting timber producing countries, including promoting equitable and fair solutions to the illegal logging problem;
2. promoting trade in legal timber, including developing and implementing specific bilateral agreements between the EU and timber producing countries;
3. promoting public procurement policies, including guidance on how to deal with legality when specifying timber in procurement procedures;
4. supporting private sector initiatives, including encouraging voluntary codes of conduct for private companies sourcing timber;
5. safeguarding financing and investment, including encouraging financial institutions investing in the forest sector to develop due care procedures;
6. using existing and/or amending legislation to support the Action Plan;
7. addressing the problem of conflict timber, including supporting the development of an international definition of conflict timber.

Of the key action areas listed in Box 2.1, two play prominent roles: the development of bilateral agreements and the setting of additional EU legislation.

Regulation (EC) 2173/2005 defined requirements for the establishment of a FLEGT licensing scheme for imports of timber into the European

Community. Under this scheme, FLEGT Voluntary Partnership Agreements (VPAs) are formed between the EU and various non-EU timber producing countries. Once a VPA is formed, trade licenses are issued, allowing for export of wood products from the producer country to the EU market. Whilst primarily aimed at guaranteeing that wood imports entering the EU derive from legal sources, VPAs are designed to achieve this through helping partner countries to enhance domestic forest governance. As suggested by their name, VPAs are entered into on a voluntary basis. Nevertheless, once entered into force, agreements become binding on both parties. The process for developing a VPA includes four phases (Table 2.1) and is intended to provide a multitude of stakeholders, from civil society and the private sector, the chance to be actively involved in developing national legality standards through consensual decision-making. At the core of each VPA is a Timber Legality Assurance System (TLAS) – a traceability scheme for wood products – which allows verification that wood is sourced and processed legally. Under a TLAS, the appointed national authority issues a FLEGT license for each verified shipment. Only materials accompanied by a valid FLEGT license can enter the EU market. The EU and its partners provide technical and financial support to an exporting country during development of a TLAS. A TLAS shall include the following five elements (Figure 2.1):

- **Legality definition** for legally produced timber. In addition, a list of the legality requirements and methods of verification required to check compliance.
- **A tracking system** that allows timber traceability from the forest to the point of export, and ensures segregation of wood from unverified and potentially illegal sources.
- **A legality verification system,** to demonstrate compliance with all applicable requirements of the legality definition and controls for the supply chain.
- **A licensing system** for wood products to be exported to the EU market; the system is executed by an appointed national authority, which issues licenses based on verification reports and other evidence.
- **A third-party independent monitoring system,** to provide evidence to interested parties that the TLAS is implemented correctly and reliably.

VPAs are currently being implemented in six countries, while another nine are in the negotiation phase (Figure 2.2). As of January 2018, FLEGT licenses have been issued only in Indonesia (started in November 2016). Ghana is expected to deliver the first FLEGT-licensed timber to the EU during 2018.

The EU FLEGT Action Plan, and VPAs in particular, are supply-side measures to tackle forest illegality. However, the Action Plan also includes the setting of additional legislation and, in particular, the development of the EU Timber Regulation (EUTR). The EUTR is one of many demand-side

*Table 2.1* Phases in a VPA development process

| Phase | Description |
|---|---|
| Phase 1<br>Pre-negotiation | The government of a timber producing country expresses interest in a VPA and the EU and its partners provide information. Based on the information received, the government discusses with the private sector and civil society and decides whether to proceed with a VPA or not.<br>This phase ends when the government of the country and the EU agree to launch formal VPA negotiations or when the former decides not to proceed with a VPA. |
| Phase 2<br>Negotiation | The EU and the negotiating country discuss the content of the VPA, with particular regard to details of the legality assurance system (LAS) and the forest governance commitments. The negotiating country sets up and implements a multi-stakeholder consultation process. Inputs from this process are discussed during negotiation with the EU.<br>Once VPA contents have been agreed, negotiations are concluded and the VPA can start. The negotiating country formally becomes a partner country. |
| Phase 3<br>Implementation | The VPA is ratified by the legislative bodies of both sides. For this purpose, it is translated into the official languages of the EU and signed by the Council of the EU, the European Commission and the legislature of the partner country. The European Parliament then approves the agreement, which can then be published. The partner country starts developing the LAS and other systems agreed in the VPA. An independent auditor checks that the LAS and the verification systems are working correctly.<br>System oversight and dispute resolution are under responsibility of a Joint Implementation Committee (JRC), which includes representatives from the partner country and the EU. |
| Phase 4<br>Licensing | Each load of wood or wood products from the partner country to the EU shall be shipped together with a FLEGT license.<br>If a FLEGT license is not issued, products cannot be placed on the EU market and are rejected at the EU border. |

Source: Own elaboration

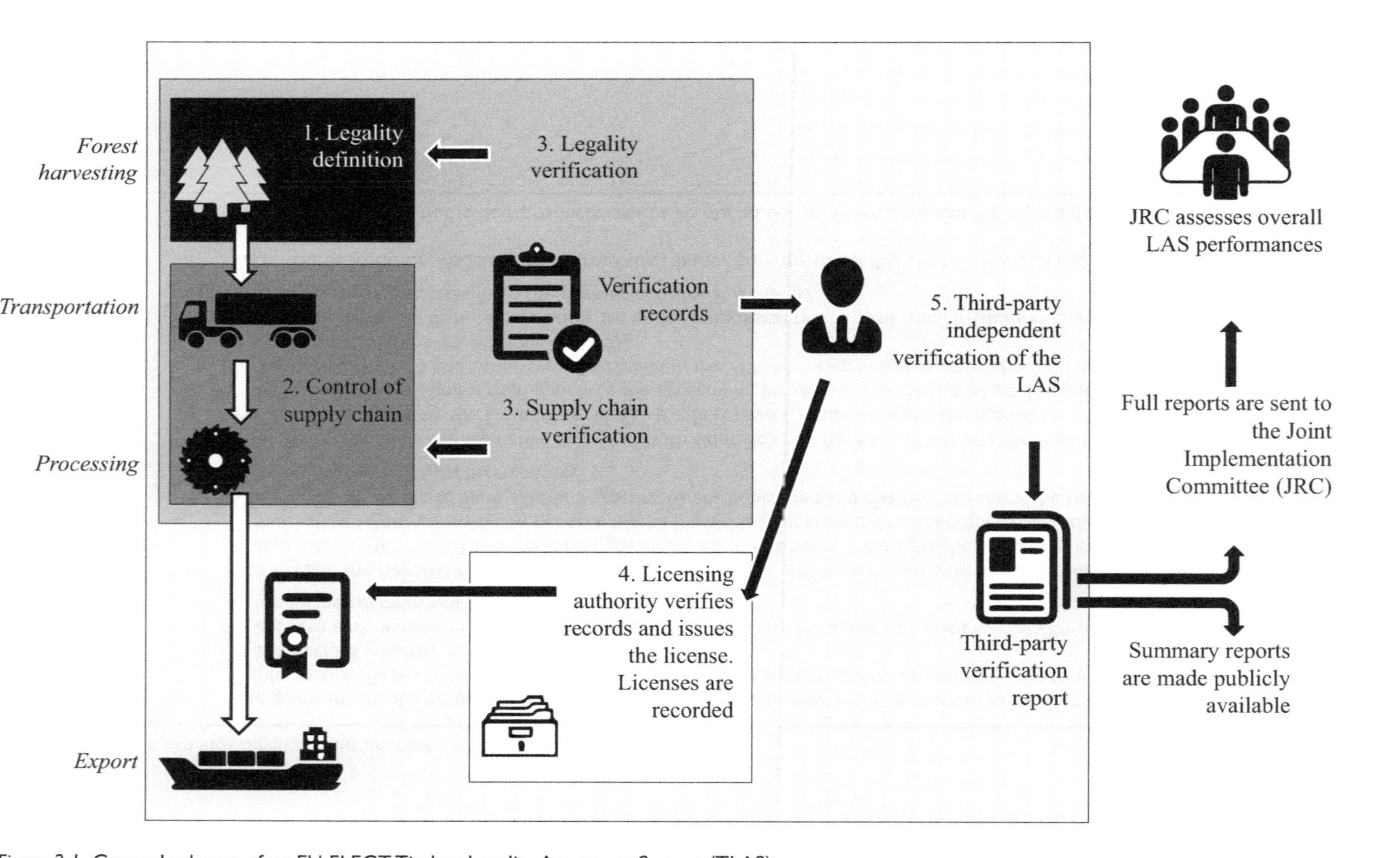


*Figure 2.1* General scheme of an EU FLEGT Timber Legality Assurance System (TLAS)

Source: Modified from EC (2007)

*Figure 2.2* Countries that are implementing or negotiating VPAs with the EU

Source: Modified from EU FLEGT Facility (2018)

initiatives for preventing trade in illegally sourced timber (which will be presented in paragraph 2.3 below).

## 2.3 Regional and national initiatives on legality

Specific initiatives for avoiding the placing of illegal timber on domestic markets have been recently developed in at least three regions/countries: the EU, United States of America and Australia. They are briefly presented below, whilst Table 2.2 presents a comparative summary of their key components. Similar initiatives are being developed in other countries, such as Japan – where the Japanese Forestry Agency has an internal action plan (2012) aiming to further promote legally verified wood-based products[1] (Momii, 2014) – and New Zealand.

### *2.3.1 The EU Timber Regulation*

Regulation (EU) 995/2010, also known as the EUTR, was passed by the EU Parliament in October 2010 and came into effect on 3 March 2013. This regulation is binding on all EU member states and sets out the obligations for all organizations within those states intending to trade timber and timber products on the European market. The ultimate goal of the EUTR is to prevent illegally harvested timber from entering the EU market.

Illegally harvested material is defined as that which is "harvested in contravention of the applicable legislation in the country of harvest" (EC, 2016). Such "applicable legislation" includes laws relating to rights to harvest timber, payments for harvesting rights and timber, environmental and

*Table 2.2* Comparative summary of Lacey Act, EUTR and Australian Illegal Logging Prohibition Act

| *Issues* | *Lacey Act* | *EUTR* | *Australian Illegal Logging Prohibition Act* |
|---|---|---|---|
| Full official name | Lacey Act (16 U.S.C. §3371 – §3378)<br>Food, Conservation and Energy Act of 2008 | Regulation (EU) 995/2010 of the European Parliament and of the Council of 20 October 2010 laying down the obligations of operators who place timber and timber products on the market<br>Delegated Regulation (EU) 363/2012 and Implementing Regulation (EU) 607/2012 | Illegal Logging Prohibition Act 2012, Act No. 166, 2012<br>Illegal Logging Prohibition Amendment Regulation 2013 (No. 1), Select Legislative Instrument No. 88, 2013 |
| Approved/entered into force | 2008/2008 | 2010/2013 | 2012/2014 |
| Scope (products covered with reference to the corresponding combined nomenclature codes) | All wood and wood-based products (including pulp and paper), except for:<br>• common cultivars (except trees), common food crops;<br>• live plants that are to remain or be planted or<br>replanted;<br>• scientific specimens of plant genetic material to be used for research, except for CITES and US nationally protected species | • Chapter 44 (4401, 4403, 4406 to 4416, 4418);<br>• Chapter 47[b];<br>• Chapter 48[b];<br>• Chapter 94 (9403 30, 9403 40, 9403 50 00, 9403 60, 9403 90 30 and 9406 00 20) | • Chapter 44 (4403, 4407, 4408, 4409 10, 4409 29, 4410, 4411, 4412, 4413 00, 4414 00, 4416 00, 4418)<br>• Chapter 47 (4701 00, 4702 00, 4703, 4704, 4705 00)<br>• Chapter 48 (4801, 4802, 4803, 4804, 4805, 4806 20, 4806 30, 4806 40, 4807 00, 4808, 4809, 4810, 4811, 4813, 4816, 4817, 4818, 4819, 4820, 4821, 4823)<br>• Chapter 94 (9401 61, 9401 69, 9403 30, 9403 40, 9403 50, 9403 60, 9403 90, 9406 00) |

| | | | |
|---|---|---|---|
| Illegal timber and applicable legislation | Timber "taken, possessed, transported or sold in violation of any law or regulation of any State, or any foreign law, that protects plants or that regulates":<br>• Theft of plants,<br>• Taking plants from officially protected or otherwise designated areas,<br>• Harvesting without required authorization, failure to pay royalties, fees or taxes related to plant's harvest, transport or trade, and<br>• Laws governing export or trans-shipment (e.g. log-export bans). | Timber harvested in contravention of the applicable legislation in the country of harvest, including:<br>• Rights to harvest timber within legally gazetted boundaries,<br>• Payments for harvest rights and timber including duties related to timber harvesting,<br>• Timber harvesting, including environmental and forest legislation, including forest management and biodiversity conservation, where directly related to timber harvesting,<br>• Third parties' legal rights concerning use and tenure that are affected by timber harvesting, and<br>• Trade and customs, in so far as the forest sector is concerned. | Timber "harvested in contravention of laws in force in the place (whether or not in Australia) where the timber was harvested", including:<br>• Legal rights to harvest;<br>• Taxes and fees related to harvesting;<br>• Timber harvesting laws, including<br>forest management and biodiversity conservation laws;<br>• Respect for third parties' legal rights and tenure;<br>• Trade and customs laws. |
| Main actors | • Animal and Plant Health Inspection Service,<br>• US Fish and Wildlife Service,<br>• US Customs and Border Protection,<br>• US Department of Justice,<br>• Importers,<br>• Traders. | • European Commission,<br>• Operators,<br>• Traders,<br>• Competent Authorities.<br>• Monitoring organizations. | • Department of Agriculture,<br>• Inspectors,<br>• Customs Minister,<br>• Importers,<br>• Processors. |

(*Continued*)

*Table 2.2* (Continued)

| *Issues* | *Lacey Act* | *EUTR* | *Australian Illegal Logging Prohibition Act* |
|---|---|---|---|
| Main obligations | Plant and Plant Product Declaration<br>Due Care System | Due Diligence System (operators)<br>Traceability (traders) | Customs Declaration<br>Due Diligence System |
| Maximum penalties | • USD$500,000 fine,<br>• Fine of double the gain/loss from transaction,<br>• Seizure of goods,<br>• Imprisonment up to 5 years. | To be defined by each EU member state and may include:<br>• Fines,<br>• Seizure of goods,<br>• Immediate suspension of authorization to trade. | • AUD$85,000 fine for individuals,<br>• AUD$425,000 fine for companies,<br>• Seizure of goods,<br>• Imprisonment up to 5 years. |

[a] Except 4412.9906 and 4412.9957
[b] Except bamboo-based and recovered (waste and scrap) products

Source: Own elaboration

forest regulation directly related to timber harvesting, third parties' legal tenure and use rights affected by timber harvesting and trade and customs regulations concerning the forest sector.

The EUTR categorizes commercial organizations involved in timber supply chains as either "Operators" or "Traders". An Operator is any organization that places timber or timber products on the internal market for the first time. This includes organizations harvesting timber from within the EU and placing it on the internal market as well as organizations importing timber or timber products from outside the EU and placing them on the market. Traders are organizations that buy, or use commercially, timber or timber products that have already been placed on the EU market. In addition to Operators and Traders, a number of other actors are involved in implementation of the EUTR. Each EU member state is obliged to assign one or more Competent Authority, responsible for the application of the EUTR within that member state. Monitoring organizations (MOs) are entities evaluated by the European Commission and found competent to assist Operators in delivery of their obligations under the EUTR. MOs are obliged to create and maintain an EUTR-compliant Due Diligence System, which can be utilized by operators and to conduct regular evaluations of any Operator that chooses to appoint an MO to oversee it. Lastly, the European Commission is charged with formulating and implementing regulations as well as evaluating and appointing MOs.

The obligations placed on Operators and Traders form the core of the EUTR. The obligations of Operators are threefold. Firstly, Operators are prohibited from placing illegally harvested timber or timber products on the market. Secondly, Operators are obliged to implement a Due Diligence System (DDS) to evaluate the risk of timber products being illegally harvested. And lastly, Operators must maintain and regularly evaluate their DDS to ensure that it is fit for purpose. The obligations placed on Traders are less onerous, requiring only that they are able to identify their suppliers and buyers, maintain such information for five years and provide this information to Competent Authorities upon request.

The due diligence process involves three main steps: collection of information on supply chains, assessment of the risk of illegality in supply chains based on that information and (if necessary) mitigation of any detected risk. The information utilized may include data on timber species, timber origin (country, sub-national region and/or concession of harvest), volumes, supplier details, documents indicating legality, etc. This information is used by the Operator to assess the risk of illegal harvesting related to the origin, species and supply chain of the applicable products and a risk conclusion is assigned. This risk conclusion is either classed as "negligible" or "non-negligible". If the risk is classed as non-negligible, the Operator must implement mitigating actions to address this risk. These may include collecting additional information to support a classification of negligible risk, conducting supplier or forest-level verification audits, sourcing certified or verified materials and/or replacing suppliers or supply chains.

The EUTR requires organizations importing timber products to the EU to investigate their supply chains with an unprecedented level of detail and scrutiny. It represents a significant demand-side market force, which complements the supply-side impetus provided by the FLEGT process. It also addresses the most fundamental problem inherent to the FLEGT process – that, after years of negotiations, only one FLEGT license has been issued. This is proving to be a double-edged sword, with the EUTR simultaneously salvaging the reputation and undermining the development of the FLEGT VPA process. The EUTR also covers a well-defined set of products compared to FLEGT VPAs, which vary from country to country. Nonetheless, major gaps in the product scope of the EUTR have been highlighted. In 2015, WWF-UK concluded that only 47% of the EU's combined nomenclature headings and sub-headings[2] that contain wood products are included within the scope of the regulation. Only 2% are officially exempt, with the remaining 51% of products falling outside the scope of the regulation. This effectively provides a legal loophole through which vast quantities of illegal wood may be legally placed on the EU market. This group includes products such as seats, printed materials and musical instruments. A tentative revision of the EUTR carried out in 2015 was expected to extend the product scope to partly address this issue, but it was not successful (WWF-UK, 2015). More recently, in the EUTR Implementation Report (EC, 2016), the European Commission stated to consider the expansion of the product scope of the regulation, and in January 2018 it launched an official public consultation on the EUTR Product Scope based on three main policy options (no change in the product scope, change by adding some products that contain timber and change by including all products that contain timber). However, the main problem to date has been the slow and uneven implementation and enforcement of the EUTR during its first years; that still remains incomplete, although substantial progress has been made recently (EC, 2016). The significant gaps among EU member states in assigning competent authorities, defining penalties and implementing checks on Operators has allowed for potential avoidance of the regulation through importation of goods into the EU via poorly performing member states, entirely undermining the legislation.

### *2.3.2 The Lacey Act*

The Lacey Act is a USA law dating back to 1900 that was amended by the USA Congress through the Food, Conservation and Energy Act of 2008 (May 2008). As a result of this amendment, plant products – including wood – were included, thus turning the Act into the world's first ban on the trade of illegally sourced wood and wood products. Under the amended Lacey Act, it is unlawful to import, export, transport, sell, receive, acquire or purchase – both in domestic interstate or foreign trade – plants and plant products in violation of any law or regulation of any USA state or any foreign law. In particular, reference shall be made to laws regarding theft of plants, taking plants from officially protected or otherwise designated areas, harvesting without required authorization, failure to pay royalties, fees or taxes related to plant's harvest, transport or trade and laws governing export or trans-shipment (e.g. log-export bans). Operators willing to import plants into the USA shall fill and provide a Plant and Plant Product Declaration that includes information about the material being imported (scientific name of the species used, country of harvest, quantity and measure and value).[3] Furthermore, operators are required to exercise "due care" in identifying the source of their plants and ensuring that they were legally harvested. Due care has been defined as the "degree of care at which a reasonably prudent person would exercise under the same or similar circumstances. As a result, it is applied differently to different categories of persons with varying degrees of knowledge and responsibility".[4] The Lacey Act, however, does not spell out what importers have to do to meet this standard. Given a certain rate of uncertainty around this concept, it might be advisable for enterprises to implement a wide range of tools and resources, including company policies and tracking procedures, questionnaires to suppliers, record keeping systems, etc. (EIA, 2009; Arnold & Porter LLP, 2012). Legality verification systems as well as forest certification might be used as part of a Due Care System, but cannot be used as stand-alone elements to comply with the Lacey Act. The Act is a fact-based, and not a document-based, statute; this means that there are no documents, licenses, marks, etc. that are accepted as a final proof of legality and it remains up to each single enterprise to the best of their knowledge to avoid buying illegal timber, depending on its profile, suppliers and sources (Forest Legality Alliance, 2015). Notwithstanding different wording, it can be assumed that in practice the steps undertaken to set up and implement a Due Care System for the purposes of the Lacey Act are not very far from those required by a Due Diligence System under the EUTR (see 2.3.1).

As regards monitoring and controls, the US Department of Agriculture's Animal and Plant Health Inspection Service (APHIS), traditionally responsible for plant imports including CITES controls, is in charge of processing Plant and Plant Product Declarations. APHIS has also the responsibility to investigate suspicious cases together with the US Department of the Interior's Fish and Wildlife Service (FWS). Support is provided by the Department of Homeland Security, which controls USA customs and monitors the borders through Customs and Border Protection. In cases where there is sufficient evidence that a plant or plant product is illegal, the case may be referred to the Department of Justice and/or forfeiture proceedings may be started.

### 2.3.3 The Australian Illegal Logging Prohibition Act

The Illegal Logging Prohibition Act of Australia came into effect in November 2012. This act made it a criminal offence to "intentionally, knowingly or recklessly import or process illegally logged timber or timber products" (ADA, 2014). This was followed by the Illegal Logging Prohibition Amendment Regulation, which came into effect on 30 November 2014, and which sets out due diligence obligations placed upon businesses.[5]

Like the EUTR and Lacey Act, the AILPA aims to prevent both importation of illegally harvested wood from outside Australia as well as the processing of any domestically grown logs that are found to have been illegally logged. The definition of illegally logged timber is also similar to the definition used under the EUTR, focusing on timber harvested in contravention of applicable legislation in the place of harvest.

The AILPA outlines the obligations of two types of actors in supply chains, "Importers" and "Processors". Importers are persons or organizations that import timber or timber products into Australia; Processors are persons or organizations that process raw logs for the purpose of trade and commerce. The law also sets out the role of "Inspectors", charged with monitoring implementation of the legislation by Importers and Processors.

The due diligence process is considered as a four-step process. The first step involves collection of information regarding the timber or timber product being imported or processed. Secondly, and as an optional step, importers/ processors may use a Timber Legality Framework and/or Country/ State Specific Guideline. These two options are tools designed to help Importers/ Processors quickly assess the associated risk of timber products. A number of Timber Legality Frameworks are listed in Schedule 2 of the regulation. These include the FSC® Forest Management Certification scheme, PEFC Sustainable Forest Management Certification and the EU FLEGT licensing scheme. Importers/Processors are then obliged to demonstrate that the Timber Legality Framework applies to the timber in their products or in the area in which the timber was harvested. If this link is demonstrated and the risk is concluded as low, the Importer/Processor is not obliged to conduct further steps in the due diligence process. Similarly, organizations may use Country or State Specific Guidelines, which provide a list of applicable legislation in the place of harvest to assist organizations in their due diligence requirements. These guidelines are also described in Schedule 2 of the regulation, however, at the time of writing, neither Country nor State Specific Guidelines were listed. Step 3 involves risk assessment of the timber or timber products intended to be imported or processed. Risk assessment under the AILPA should consider both the risk associated with the species and the origin of timber. The risk assessment process must identify the risk level of products, either as "no risk" or "risk". If "risk" is concluded and this cannot be concluded as "low risk", then mitigating actions must be implemented as part of step 4.

As a result of the AILPA being developed later than the Lacey Act and EUTR, it draws on different themes from both. Like the Lacey Act, a declaration of timber legality is required for each shipment at the point of import (which is not required under the EUTR). As with the Lacey Act, penalties can result in a maximum financial penalty or incarceration up to 5 years. This is in contrast to the EUTR, which devolves decisions on penalties to individual member states. The AILPA does not enshrine a role for organizations assisting Importers/Processors with their due diligence (the role of MOs under the EUTR), much like the Lacey Act. However, when it comes to product scope, the AILPA is more similar to the EUTR, covering products such as particle board, fibreboard, pulp and paper and furniture, which are not covered by the Lacey Act.

The AILPA also has a unique approach to forest certification schemes, compared to the EUTR and Lacey Act, in that such voluntary schemes are accepted as key indicators of low risk. An Importer need only gather information about certified products, confirm that the certification scheme is a recognized Legality Framework under the AILPA and confirm that the product certification is legitimate. Importers must also take into account "information they know or ought reasonably to know" about the legality of the product, but if they still conclude the risk to be low, full risk assessment is not required. This approach is distinct from the EUTR and Lacey Act, which simply approach certification as one of many potential tools that can be used to indicate risk.

## 2.4 Final considerations

More than 20 years after the first discussions by domestic and intergovernmental bodies surrounding illegal logging, we are beginning to see robust and effective results. The growing number of similar policies indicates that the first initiatives implemented by the USA and the EU are being imitated by other governments, setting the bar for legislative best practice. Whilst they can contribute to addressing illegal logging issues, some risks and limitations can be identified regarding their effectiveness and efficiency, and in terms of potential induced market diversions and trade-offs with voluntary tools.

It should be noted that 12 years after the launching of the FLEGT Action Plan, only six VPAs have been finalized and only one FLEGT license issued. Many key producer countries still remain outside the FLEGT VPA process, and in some cases (e.g. Brazil, India and China) will likely never enter. VPAs often require specific cooperation agreements to be established, for example the Bilateral Coordination Mechanism (BCM) between the EU and China.

Moreover, with only the exception of the Central African Republic (FERN, 2013), VPAs are intended to cover all exports, including those to non-EU markets and trade within domestic markets. However, implementation efforts so far have been focused on EU-oriented exports and related issues. Domestic markets are largely informal and mostly consist of chainsaw logging and milling. The contribution to local socio-economic conditions can be huge. In countries like Ghana (Pearce quoted by FERN, 2013) or Cameroon (Pye-Smith, 2010; Cerutti and Lescuyer, 2011), the economic dimension can be higher than that officially registered for the industrial sector.

Whilst the VPA process has the potential for immense benefits, prominent NGOs have also raised grave concerns regarding their implementation. For example, in 2014 the Anti-Forest Mafia Coalition criticized Indonesia's TLAS, named Sistem Verifikasi Legalitas Kayu (SVLK), on a number of counts. These included SVLK's use of a poor legality definition, low credibility of the SVLK certification standard, weak independent monitoring and certification of operations that may potentially source illegal timber.

At the macro-level, the implementation of legality requirements may affect market trends and trade patterns. Legality requirements do not cover wood trade at the global level and thus may, for example, drive the diversion of wood exports to destinations with less stringent regulatory frameworks. Indeed, part of the wood that is imported and locally consumed by emerging economies (such as India and China) is subject to low or no requirements, and may feed informal or illegal harvesting. The final potential consequence of these market- and policy-driven dynamics within the international forest regime may be the creation of a dual market, with some trade flows under stricter constraints than others. Giurca et al. (2013), for example, reported that producer countries in Asia, like Indonesia and Malaysia, were increasing their exports to other regional markets in Asia and the Middle East because they adhere to less stringent regulations and provide continuity of trade flows. Additional side effects of legality requirements implemented by certain countries may include substitution effects and impacts on small and medium sized enterprises (SMEs). Substitution may occur between wood species or products from high-risk sources (e.g., tropical countries) with others from low-risk sources (e.g., temperate ones). For example, the UNECE/FAO (2014) reported that in the last few years oak has increased its dominant market position in the European flooring and joinery sectors, whilst tropical hardwoods have continued to lose market share. The ITTO (2012) reported a continuous trend towards the substitution of tropical plywood with softwood and temperate hardwood plywood and other panels. Substitution effects may also include shifts from wood-based products to different materials (e.g., plastics or wood-plastic composites) especially in the building sector (windows, decking, etc.).

SMEs often do not have the knowledge and/or resources to address legality policies, especially bearing the costs of implementing traceability procedures. As a consequence, they might decide either to stop exporting towards markets with strict legality requirements or to sell their products to bigger exporters. In a similar way, small and medium EU or US importers may not be able to maintain due diligence or Due Care Systems and thus be induced to stop importing timber from risky sources or simply buy from larger importers (Karsenty et al., 2014).

Finally, potential trade-offs between compulsory and voluntary initiatives should be taken into consideration as well. The main international forest certification schemes have recently reviewed and updated their standards to address many issues imposed by legality requirements. The role of forest certification under legality requirements, however, remains unclear. Although certification is amongst the tools companies may use to assess and mitigate risk, certified materials are not given a green lane when entering the EU or US markets. This could lead to a shift in perceptions of the importance of certification in comparison to compulsory tools, pushing companies to concentrate their efforts and limited resources on obligatory legality requirements, rather than investing in voluntary certification. On the other hand, certification can help to reduce risk in supply chains, and may prove more cost-effective than on-site supplier audits or than replacing entire supply chains. A recent study performed by Karsenty et al. (2014) in Cameroon, Gabon and the Republic of the Congo suggests that trade statistics do not indicate a decrease in certified tropical wood products being placed on the EU market. There is also no significant decrease in the number of certificates in VPA countries or the main importing countries. Thus, it is expected that forest certification could benefit from the introduction of legality requirements around the world. At the moment, these legality tools are still in their infancy and enough time has not yet passed to form an accurate picture of the effects they have on forest certification.

## Notes

1 Japan has a voluntary system to promote the sourcing of legally verified wood-based products – the *goho*-wood system – which is coordinated by the Japan Federation of Wood Industry Associations (JFWIA) and which the private sector is encouraged to follow (Momii, 2014). For further information: http://goho-wood.jp/world/.
2 All products declared to EU customs are classified using the combined nomenclature (CN) headings, a classification system used to determine the rate of customs duty. These CN codes are also used to delineate the product scope of the EUTR (see Annex 1 of Regulation (EU) 995 2010).
3 Declarations shall be done using a specific form and only with regard to scheduled product categories (e.g., pulp and paper products are currently not included). A free online resource for filling and submitting declarations, called Lacey Act Web Governance System (LAWGS), has been developed and is available at https://lawgs.aphis.usda.gov/lawgs/.
4 Senate Report No. 97–123 (Comm. on Environment and Public Works).
5 Henceforth these two pieces of legislation will be referred to collectively as the Australian Illegal Logging Prohibition Act (AILPA).

## References

ADA (Australian Department of Agriculture) (2014). Illegal Logging. Updated 4 December 2014. www.agriculture.gov.au/forestry/policies/illegal-logging.

Arnold & Porter LLP (2012). Interpreting the Lacey Act's "Due Care" Standard after the Settlement of the Gibson Guitar Environmental Enforcement Case. www.arnoldporter.com.

Cerutti, P.O., Lescuyer, G. (2011). The Domestic Market for Small-Scale Chainsaw Milling in Cameroon: Present Situation, Opportunities and Challenges. Occasional Paper 61. CIFOR, Bogor.

EC (2003). Forest Law Enforcement, Governance and Trade (FLEGT). Proposal for an EU Action Plan. COM (2003) 251. Commission of the European Communities, Brussels.

EC (2007). Forest Law Enforcement Governance and Trade. Guidelines for Independent Monitoring. FLEGT Briefing Notes N. 7. European Commission, Brussels.

EC (2016). Report from the Commission to the European Parliament and the Council. Regulation EU/995/2010 of the European Parliament and of the Council of 20 October 2010, Laying Down the Obligations of Operators Who Place Timber and Timber Products on the Market (the EU Timber Regulation). http://ec.europa.eu/environment/forests/eutr report.htm.

EIA (2009). The U.S. Lacey Act. Frequently Asked Questions about the World's First Ban on Trade in Illegal Wood. Version II. Updated January 2009. www.eia-global.org/lacey.

EU FLEGT Facility (2018). VPA Countries. www.euflegt.efi.int/vpa-countries.

FERN (2013). *Improving Forest Governance: A Comparison of FLEGT VPAs and Their Impact*. FERN, Brussels.

Forest Legality Alliance (2015). All You Need to Know about the US Lacey Act, the EU Timber Regulation and the Australian Illegal Logging Prohibition Bill. International Developments in Trade in Legal Timber. www.forestlegality.org.

G8 (2002). G8 Action Programme on Forests Final Report. www.illegallogging.info.

Giurca, A., Jonsson, R., Rinaldi, F., Priyadi, H. (2013). Ambiguity in Timber Trade Regarding Efforts to Combat Illegal Logging: Potential Impacts on Trade between South-East Asia and Europe. *Forests*, 4: 730–750.

ITTO (2012). Annual Review and Assessment of the World Timber Situation 2012. www.itto.int/annual_review/.

Karsenty, A., Lemaître, S., Dessard, H. (2014). *Potential Causes of the Contraction of the Demand for FSC Certified Tropical Timber in the European Union*. CIRAD, Montpellier.

Momii, M. (2014). *Trade in Illegal Timber: The Response in Japan: A Chatham House Assessment*. Chatham House, London.

Pye-Smith, C. (2010). *Cameroon's Hidden Harvest*. CIFOR, Bogor.

UNECE/FAO (2014). *Forest Products Annual Market Review 2013–2014*. United Nations, Geneva.

WWF (2014). FLEGT Gaps, Opportunities and Asks. http://barometer.wwf.org.uk/what_we_do/government_barometer/what_is_flegt_/the_flegt_process/flegt_gaps_opportunities_asks/.

WWF-UK (2015). *Can the European Union's Timber Regulation Keep Out Illegal Timber?* WWF, London.

Chapter 3

# Corporate governance and related voluntary tools

## Codes of conduct, sustainability reporting, compensation and commitments

*Aynur Mammadova and Davide Pettenella*

This chapter aims at providing an overview of corporate governance and related voluntary tools. After introducing the background on the emergence of voluntary tools and the willingness towards self-regulation by businesses, the chapter then describes codes of conduct (paragraph 3.2), corporate reporting (paragraph 3.3), compensation and offsetting (paragraph 3.4) and commitments (paragraph 3.5). Particular focus is given to commitment types and characteristics, implementation strategies and shortcomings and challenges.

### 3.1 Background: the emergence of voluntary tools

With the widespread acceptance of "market-based" solutions to environmental problems and the rise of the neoliberal approach to forest governance, businesses have emerged as important players in the arena. By considering all strict "command and control" measures undesirable and redundant, corporations have been more willing to demonstrate their progress by adopting different voluntary sustainability strategies or by participating in sector-wide roundtables and standard-making. This willingness towards self-regulation has manifested in widespread adoption of corporate norms, codes of conduct and voluntary standards, leading to a more complex network of global environmental governance (Keohane and Nye 2000).

Since the adoption of the UN Global Compact Ten Principles, corporate social responsibility (CSR) has gradually become part of corporate strategy (Humphreys 2012). Lately, forests and landscapes have gained centrality in CSR, whether it is about calculating and mitigating greenhouse gas emissions, enhancing freshwater provision, protecting biodiversity or investing in community development.

An increasing number of businesses also recognize forests and deforestation as an important business risk factor when it comes to reputational risks, regulatory action, brand equity and consequent financial returns (Ceres 2017). Among the 201 companies that have disclosed their data and were analysed within the 2017 CDP (formerly Carbon Disclosure Project) report, around 87% identify deforestation as a business risk and around one-third (32%) report that they are already facing impacts on their businesses (CDP 2017). Thus, with the recognition that forest-related risks and opportunities affect availability of resources, core business and resilience of the companies, the topic is no longer limited to the scope of individual CSR or marketing departments.

## 3.2 Codes of conduct

Codes of conduct or codes of ethics can be described as early generation of corporate self-regulation. Kaptein (2004) defines codes of conduct as aspirational instruments to enhance social responsibility, while according to Nijhof et al. (2003) codes of conduct contain open guidelines describing desirable behaviours and closed guidelines prohibiting certain behaviours. In general, a code of conduct can be defined as a written document articulating the values and ethical principles of the organization. Starting from the 1980s and 1990s, adopting corporate ethical codes trended among big US corporations, later extending to Europe and other geographies as well. The main premises of adopting codes of conduct were to prevent an employee misbehaviour and illegality leading to profit reductions or legal accusations and fines (Adams et al. 2001). By that time, codes of conduct developed to include more thematic areas such as product safety, ethics and indiscrimination at the workplace, child and forced labour, environment, etc. Usually the scope of application of codes of conduct extends from employee behaviours to requirements towards outsourcing companies or suppliers. The content of a typical corporate codes of conduct would include an introduction explaining the scope and affected parties, considerations on legal compliance, internal health and safety regulations, labour rights, anti-corruption, supply chain management and environment (Stevens 2017). Each one of these sections would refer to relevant international conventions, guidelines and documents, i.e., International Labour Organization (ILO) conventions or EU Waste Regulation.

There is a considerable amount of literature trying to evaluate the effectiveness of corporate codes and their impact on employee behaviour, consumer choice and overall performance of the companies (Kaptein 2004; Stevens 2008; Stevens 2017). Unfortunately, as self-proclaimed policies, the quality of observance and actual realization of the codes of ethics in the everyday life of a company largely depend on the company's self-assessment and internal audits. Due to this reason, many see the risk for conflict of interest, ungrounded claims and greenwashing when it comes to actual observance of codes of ethics. However, there are few audit schemes and joint initiatives, such as SEDEX or the Business Social Compliance Initiative (BSCI), that help to bring some objectivity to the assessments (Nijhof et al. 2003). Despite lacking legal enforcement mechanisms, these instruments also help civil society to hold businesses accountable against their own written principles.

When it comes to the specific topic of forests, corporate codes mainly focus on aspirations, policies and desired actions to reduce the deforestation impact of their operations. Based on the sector and already existing sustainability standards in that specific sector, the company codes can include sourcing from suppliers with no deforestation record "post-(the sector-wide agreed baseline year)" or to source only from certified suppliers. The "no post-2006 deforestation" policy in accordance with the Soy Moratorium in Brazil, which is also manifested in Mars or Kellogg's Global Supplier Codes of Conduct, can serve as an example for this (Kellogg's 2015; Mars Inc. 2018).

The "private business clubs" such as the Consumer Goods Forum (CGF)[1] and World Business Council for Sustainable Development (WBCSD)[2] can also help mobilize the resources, define best practices and set sector-wide norms as aspirations for individual corporate codes of conduct around the topic of forests and deforestation (CGF 2018; WBCSD 2018).

## 3.3 Corporate reporting

While traditionally corporate codes of conduct direct attention to internal processes and stakeholders, a next generation voluntary sustainability tool called corporate reporting focuses more on external stakeholders. Growing need for more corporate transparency made financial reporting alone insufficient for meeting the expectations of diverse stakeholders, mainly of investors and consumers. Among different CSR tools, corporate reporting has emerged as a mechanism to communicate the progress towards set goals and aspirations, bring transparency and build trust with diverse stakeholders. Thus, along with financial/economic reporting, the performance and progress in social and environmental aspects have become important aspects of corporate reporting. Formerly known as ESG (environment, social and governance), reporting now is widely referred as Corporate Sustainability Reporting.

The Global Reporting Initiative (GRI) defines sustainability reporting as "the practice of measuring, disclosing and being accountable to internal and external stakeholders for organizational performance towards the goal of sustainable development" (2018). GRI itself is a nonprofit organization established in 1997 that provides guidelines and framework for corporate reporting based on environmental, social, economic and governance aspects as four main key areas of sustainability. The latest framework developed by the organization is the GRI Standards, released in 2016, offering a modular structure for reporting and making it easier for updates and adaptation. The GRI Indicator Protocol Set includes 30 environmental indicators (EN1-EN30) and offers criteria on energy, biodiversity and emissions. The environmental indicators and criteria offer great flexibility for integrating forests as an important element of corporate reporting by businesses in diverse sectors (GRI 2018).

There are numerous other international and national guidelines on how businesses are expected to report on their activities and progress towards sustainability. Besides the GRI Standards, the main international ones are:

- the Organization for Economic Cooperation and Development (OECD Guidelines for Multinational Enterprises);
- the United Nations Global Compact (the Communication on Progress);
- the International Organization for Standardization (ISO 26000, International Standard for Social Responsibility).

Paragraph 47 of "The Future We Want", the resulting document of the UN Rio+20 conference, stresses the importance of corporate sustainability reporting (Assembly UG 2012). EU Directive 95/2014 brings more stringency to the topic and makes it compulsory for businesses with a certain turnover threshold (and those with >500 employees) to produce non-financial and diversity reporting (Eur-lex.europa.eu 2014). However, the legislation leaves it to businesses to choose among the list of international and national guidelines as the most appropriate and applicable for their own reporting.

With the growing internalization of forest-related risks, the sustainability reports of businesses dealing with forest-risk commodities[3] include separate sections on supplier policies, commitments, targets and achieved progress in accordance with their impact on forest ecosystems. When it comes to consolidated business reporting on forests, the CDP Forests Program and its annual Global Forests Reports analyses risks and opportunities linked to forest-risk commodities[4] by acting on behalf of 650 signatory investors and around 200 participating companies (CDP 2017).

## 3.4 Compensation and offsetting

The idea of offsetting or compensating for environmental damage developed much later as a voluntary tool and with the advance of legislation became compulsory in some fields. The principles of the Kyoto Protocol (1997) offered an initial widespread acknowledgement of offsetting as a potential market-based tool. However, in light of the broken promises of the Protocol, the European Union Emissions Trading System (EU ETS) became the first internationally operating carbon trading scheme in 2005. EU ETS covers 31 European countries and regulates emissions from 11,000 installations. Within a "cap and trade" mechanism of the ETS, companies receive and buy allowances for their emissions and can access international credits from "carbon saving" projects around the globe (EC 2018). Within the context of the Kyoto Protocol and EU ETS, forests are viewed as carbon sinks and their role in climate mitigation strategies are gaining importance (see Chapter 8 for more details).

Forests have become a central element of the "biodiversity compensation" concept as well. In 2002, the World Summit on Sustainable Development stressed the urgency of dealing with the global biodiversity loss and a target of "a significant reduction in the rate of global biodiversity loss by 2010" was agreed as the result. The EU set a more ambitious target of achieving a full halt to biodiversity decline by 2010, which was later modified into a 2020 goal. The target focused on infrastructure and development activities that the companies are required to mitigate (avoid, minimize, restore and rehabilitate) and offset the negative impact on biodiversity to achieve no net loss or even net gain in biodiversity within the area of a development project. The implementation mechanisms are identified as one-off offset (compensation directly paid by a developer for inflicted negative impact), financial compensation (setting aside funds for biodiversity conservation or contributing to already existing project) and accessing credits from mitigation banking (credits available from previous biodiversity conservation and mitigation activities by a third party) (Bennett et al. 2017).

At an international scale, the main driving force for voluntary business actions towards biodiversity compensation is the International Finance Corporation's (IFC) Performance Standard 6 on Biodiversity Conservation and Sustainable Management of Living Natural Resources. The Business and Biodiversity Offset Program (BBOP) is yet another initiative providing a set of criteria and indicators for businesses deciding on mitigation and compensation activities, and it is recognized by IFC as a best practice for its own risk mitigation (IFC 2017).

While companies might opt for adopting biodiversity offsets voluntarily, within Europe the main drivers for their uptake have become legislation such as EU Habitats Directive (Council Directive 92/43/EEC) or EU Directive on Environmental Impact Assessment (Directive 2011/92/EU) and their integration to European-wide national legislations (Bennett et al. 2017). Similarly to natural forest areas, wildlife corridors and wetlands are considered critical habitats for biodiversity, their conservation and restoration are prioritized within the "biodiversity compensation" scheme once more.

Reducing Emissions from Deforestation and Forest Degradation in Developing Countries, in other words, REDD+ schemes, offer great flexibility and potential for synergy between carbon market, biodiversity compensation and forest conservation activities for companies that seek "no net loss" or "net gain" of biodiversity. However, robust legislation and implementation mechanisms are needed to ensure better coordination between the three and avoid any possible leakage effect (Lanius et al. 2013).

## 3.5 Commitments

While typically the focus of corporate codes of conduct and reporting is more on the internal operations and stakeholders, corporate commitments direct attention to external stakeholders and upstream suppliers. The thematic of corporate commitments can range widely from ethical issues to a (non)purchase decision of a very specific product and can be viewed as an overarching measure combining different tools mentioned earlier. Corporate commitments towards achieving a deforestation-free global economy is a relatively new development that was initially driven by civil society demands and lately by different legislation and legislative frameworks.

For many years, global demand for forest derived products and industrial agriculture has been considered the culprit of deforestation, especially in the tropics. To this end, the 2014 New York Declaration on Forests (NYDF) became an important document stressing the shared roles and responsibilities of private finance and businesses in clearing global production systems from deforestation (Summit 2014) As of today, 60 public, 59 private and 73 civil society organizations have adhered to NYDF to pledge to at least halve the rate of natural forest loss by 2020 and strive to end it by 2030 (Lambin et al. 2018).

Despite the voluntary character, the declaration together with the 2016 Paris Climate Agreement has spurred adoption of public and private deforestation-related commitments. A plethora of other initiatives have also focused on the issue as an attempt for better accountability and monitoring of those commitments. Just to name a few, Forest 500, Supply Change by Forests Trends, Consumer Goods Forum, Tropical Forest Alliance 2020, CDP Forests Program, Accountability Framework and Collaboration for Forests and Agriculture (CFA) are the third-party or multi-stakeholder initiatives emerging as the result of extended attention on the issue.

### *3.5.1 Bird's-eye view: commitment types and characteristics*

The most comprehensive dataset about deforestation-related corporate commitments is regularly presented by Supply Change Initiative of Forest Trends (Supply Change 2018). By the time of this publication, the initiative was able to track 800 companies within the "big four" commodity sectors: palm oil, soy, cattle and timber and pulp (see Figure 3.1). Of these, 469 have publicly declared their commitment(s) to addressing deforestation (Donofrio et al. 2017). By acting as a platform that presents the aggregated statistics on corporate commitments and their per sector allocation, the Supply Change does not necessarily evaluate the quality of either the commitments nor their implementation strategies. This task is undertaken by another initiative called Forest 500 by Global Canopy, which identifies and ranks 500 power brokers as the most influential companies, financial institutions and

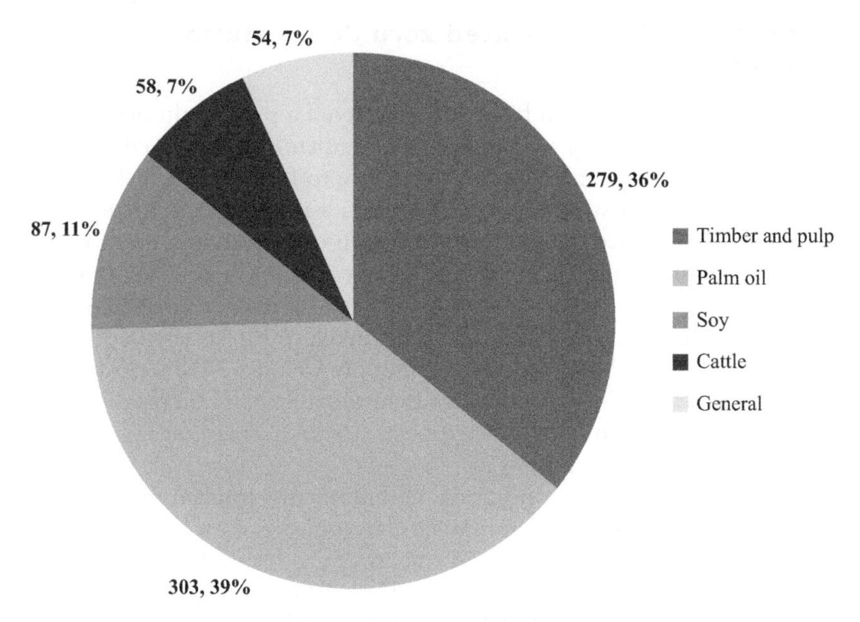


*Figure 3.1* Corporate commitment count per commodity, tracked by Supply Change

Note: The figures are presented as number of observations, percentage

Source: Donofrio et al. (2017)

governments in forest-risk commodity supply chains and attempts to hold them accountable for their actions.

The Supply Change categorizes commitments based on different factors such as product type and coverage, scope of operations, targets, timeline and policies (procurement, etc.) (Supply Change 2018). Classification based on commitment target results in the following most common types:

- zero deforestation;
- zero gross deforestation;
- zero net deforestation;
- peatland protection;
- HCV area protection;
- HCS area management/protection.

Another similar classification is provided by Lambin et al. (2018), who identifies company commitments as zero deforestation, zero net deforestation, no deforestation of valuable forests, sustainable or responsible forest management or a commitment to a net increase in forest area.

**Box 3.1 Palm-oil-related zero deforestation commitments**

For a long time, palm oil has been considered a perfect alternative to other vegetable oils: as a nutrient rich ingredient it can be used within many products ranging from personal care to food, and its relatively low production cost makes it attractive as a raw material for bioenergy. However, once the link between palm oil production and forest destruction in producing countries had been established, many public policies have been designed to avoid the negative impact. Around 60% of palm oil exports are estimated to be linked to forest destruction in Indonesia and Malaysia since 2000 (Supply Change 2018). However, as much attention is directed towards sustainable palm oil production in South-East Asian countries, tropical Africa is emerging as a new frontier for palm plantations.

Among the zero deforestation commitments tracked by Forest Trends, palm-oil-related commitments have the highest share, accounting for 303 in total. As a general rule, they are focused on purchases of RSPO certified oil and define 2020 or 2030 as a target year for achieving 100% traceability. Ferrero S.p.A and its Palm Oil Charter can serve as example for this type of commitment. According to self-reported progress, its pledge to "achieving 100% traceable segregated, RSPO-certified palm oil by the end of 2014" is already on track to create a basis for further engagement in the topic (Ferrero 2018).

### *3.5.2 Implementation strategies of commitments*

Although the forest certification schemes and their criteria are not inherently designed to deal with zero-deforestation-related topics, in the absence of suitable tools, the increasing number of businesses opt for certification and standards such as Forest Stewardship Council (FSC®) or the Roundtable on Sustainable Palm Oil (RSPO) to demonstrate some progress against their time-bound commitments. In some other cases, companies develop their own criteria, policies and supplier codes of conduct around protecting High Conservation Value (HCV), high carbon stock (HCS) areas, respect for free, prior and informed consent (FPIC) of local communities and work on developing private traceability tools of their supply chains (Donofrio et al. 2017).

### *3.5.3 Shortcomings and challenges*

Starting from Brown and Zarin (2013), who argued against the vague and generic language of the corporate commitments and translation of these commitments to a positive action, corporate pledges have received a good degree of criticism. Without proper implementation strategies and action plans, third-party monitoring and regular reporting, these commitments might be prone to greenwashing and have little to no impact on the ground. Moreover, as the Supply Change tracking also indicates, much attention has been towards palm oil, while other important deforestation-risk commodities such as cattle and soy have received less spotlight, leading to a comparatively low number of corporate commitments (Supply Change 2018; Lambin et al. 2018).

The extent of attainability vs. ambition of commitments has been discussed widely, especially when comparing Zero Net Deforestation (deforestation is allowed with a condition of offsetting it with reforestation somewhere else) with Zero Gross Deforestation (no deforestation of natural forests) commitments. While some argue that given the complexity, Zero Net Deforestation is the most attainable and practical approach to achieve the set goal for 2020, others claim that this approach has a risk of translating to very little on the ground, leading to leakage and feeding greenwashing practices further. On the other hand, although in theory Zero Gross Deforestation sounds like the most rigid option for the ambitious climate goals, reality is much more complex. Recently "no harvest regimes" have been criticized for leading to illegal cuts, discriminating against local communities and making forestry a less attractive sector for further investments.

Regardless of the wording, these commitments or corporate pledges have faced challenges in different directions. Those challenges can be grouped into design and implementation phases.

The main design-related challenges are as follows:

1 Definitions and pledges utilized by businesses are quite diverse. The corporate language could be very vague and generic, leading the way to diverse interpretation of the commitments.
2 In the absence of clear sector-wide standards, the pledges have different deforestation cutoff baseline year and target dates. This further exacerbates the reporting, monitoring and impact evaluation. Thus, harmonization and development of guidelines are necessary.

The implementation- and monitoring-related challenges are:

3 There are different actors involved in different nodes of the global trade. Harmonizing the pledges and coordinating them across complex and globalized supply chains is a challenging task.
4 The lack of guidance for good practices make it difficult for companies to find the optimal implementation strategies that could be accepted by external stakeholders.
5 Businesses are sceptical about reporting about little progress or difficulties during the implementation phase, which can lead to "dormant" commitments, as described by Donofrio et al. (2017).

6 As tools and initiatives proliferate but do not harmonize, there is a certain level of ambiguity and confusion about the estimation of real outcomes.

However, despite all the uncertainty and ambiguity involved, corporate zero deforestation commitments present an important first step toward acknowledging the adverse impact on global forests and showing willingness to act. The frameworks by multi-stakeholder initiatives such as Collaboration on Forests and Agriculture (CFA) (WWF 2018) and Accountability Framework (2018) offer great potential for the next steps in terms of harmonizing the commitment design and implementation strategies across companies and sectors. Meanwhile, different certification schemes such as RSPO and RTRS try to address the topic by modifying their criteria and principles.

## Notes

1 CGF brings together the CEOs and senior management of over 400 retailers and manufacturers in diverse sectors across 70 countries to work towards better consumer trust (CGF 2018).
2 WBCSD is a global, CEO-led organization of over 200 leading businesses working together to accelerate the transition to a sustainable world (WBCSD 2018).
3 The term "forest risk commodity" was coined by the Global Canopy Program 2013 report, in which they define it as "globally traded goods and raw materials that originate from tropical forest ecosystems, either directly from within forest areas, or from areas previously under forest cover, whose extraction or production contributes significantly to global tropical deforestation and degradation" (Rautner et al. 2013).
4 CDP Global Forests Reports focus on businesses associated with the production and trade of timber, palm oil, cattle and soy as the four main forest-risk commodities.

## References

Accountability Framework. 2018. Accessed 28/03/2018. URL: https://accountability-framework.org/about-us/#framework

Adams, J.S., Tashchian, A., Shore, T.H. 2001. Codes of Ethics as Signals for Ethical Behavior. *Journal of Business Ethics*, 29, p. 199. https://doi.org/10.1023/A:1026576421399

Assembly, U.G. 2012. The Future We Want. *Resolution*, 66, p. 288.

Bennett, J., Chavarria, A., Ruef, F., Leonardi, A. 2017. State of European Markets 2017: Biodiversity Offsets and Compensation. *Ecostar Hub*. Available at: www.ecostarhub.com/wp-content/uploads/2017/06/State-of-European-Markets-2017-Biodiversity-Offsets-and-Compensation.pdf

Brown, S., Zarin, D. 2013. What Does Zero Deforestation Mean? *Science*, 342(6160), pp. 805–807.

CDP. 2017. From Risk to Revenue: The Investment Opportunity in Addressing Corporate Deforestation. *CDP Report 2017*. Available at: http://b8f65cb373b1b7b15feb-c70d8ead6ced550b4d987d7c03fcdd1d.r81.cf3.rackcdn.com/cms/reports/documents/000/002/860/original/CDP-2017-forests-report.pdf?1511199969

Ceres. 2017. Case Study Series: Business Risk from Deforestation. Available at: www.ceres.org/sites/default/files/Engage%20the%20Chain/ETC%20Climate%20Advisors%20Case%20Studies%20(1).pdf

Consumer Goods Forum (CGF). 2018. Who We Are: Five Strategic Initiatives.

Donofrio, S., Rothrock, P., Leonard, J. 2017. *Supply Change: Tracking Corporate Commitments to Deforestation-Free Supply Chains*. Forest Trends, Washington, DC.

Eur-lex.europa.eu. 2014. Directive 2014/95/EU of the European Parliament and of the Council of 22 October 2014 Amending Directive 2013/34/EU as Regards Disclosure of Non-Financial and Dinformation by Certain Large Undertakings and Groups Text with EEA Relevance.

European Commission (EC). 2018. Emissions Trading System (ETS).

Ferrero, S.P.A. 2018. Palm Oil Charter. Accessed 05/04/2018. Available at: www.ferrero.com/group-news/Ferrero-Palm-Oil-Charter

Global Reporting Initiative (GRI). 2018. Sustainability Reporting. Accessed 02/04/2018.

Humphreys, D. 2012. *Logjam: Deforestation and the Crisis of Global Governance*. Routledge, Abingdon, UK.

International Finance Corporation (IFC). 2017. Performance Standard 6. Available at: www.ifc.org/wps/wcm/connect/topics_ext_content/ifc_external_corporate_site/sustainability-at-ifc/policies-standards/performance-standards/ps6

Kaptein, M. 2004. Business Codes of Multinational Firms: What Do They Say? *Journal of Business Ethics*, *50*, pp. 13–31.

Kellogg's. 2015. Kellogg Company Global Supplier Code of Conduct. Available at: www.kelloggcompany.com/content/dam/kelloggcompanyus/corporate_responsibility/pdf/2015/GlobalSupplierCodeofConductResourceGuide.pdf

Keohane, R.O., Nye Jr, J.S. 2000. Globalization: What's New? What's Not? (and So What?). *Foreign Policy*, pp. 104–119.

Lambin, E.F., Gibbs, H.K., Heilmayr, R., Carlson, K.M., Fleck, L.C., Garrett, R.D., de Waroux, Y.L.P., McDermott, C.L., McLaughlin, D., Newton, P., Nolte, C. 2018. The Role of Supply-Chain Initiatives in Reducing Deforestation. *Nature Climate Change*, p. 1.

Lanius, D.R., Kiss, E., Besten, J.W. 2013. *Aligning Biodiversity Compensation and REDD+: A Primer on Integrating Private Sector Conservation Financing Schemes in the Tropics and Sub-Tropics*. IUCN NL, Amsterdam.

Mars Inc. 2018. Our Soy Sourcing and Deforestation Policy.

Nijhof, A., Cludts, S., Fisscher, O., Laan, A. 2003. Measuring the Implementation of Codes of Conduct: An Assessment Method Based on a Process Approach of the Responsible Organization. *Journal of Business Ethics*, *45*, pp. 65–78.

Rautner, M., Leggett, M., Davis, F. 2013. *The Little Book of Big Deforestation Drivers*. Global Canopy Programme, Oxford.

Stevens, B.J. 2017. Corporate Ethical Codes: Effective Instruments for Influencing Behavior. *Journal of Business Ethics*, *78*, p. 601. https://doi.org/10.1007/s10551-007-9370-z

Summit, U.C. 2014. *New York Declaration on Forests*. United Nations, New York, NY.

Supply Change. 2018. About: Commitments: Forest Trends.

World Business Council for Sustainable Development (WBCSD). 2018. Forest Solutions Group.

WWF. 2018. Collaboration for Forests and Agriculture. Available at: www.wwf.org.br/natureza_brasileira/reducao_de_impactos2/agricultura/agr_acoes_resultados/copy_of_colaboracao_para_florestas_e_agricultura__cfa___27062017_1949/

Chapter 4

# Certification schemes and processes

*Laura Secco and Mauro Masiero*

This chapter aims to provide an introduction to certification schemes and processes applicable to the forest sector. After a general introduction to the topic (paragraph 4.1), basic general concepts and terminology are introduced (paragraph 4.2) followed by an overview of the main certification schemes that might be used in forestry (paragraph 4.3) and finally some general conclusions are drawn (paragraph 4.4).

## 4.1 Introduction: certification as a tool for corporate social responsibility

The emergence of sustainable forest management (SFM) certification standards and guidelines, which occurred between the end of the 1980s and the beginning of the 1990s, has to be connected to changed paradigms related to forest management approaches and policy-making processes (Cashore et al., 2003; Bass, 2003; Lemos and Agrawal, 2006): a transition from traditional government owning monopoly on decision and policy-making authority, to procedures in which state policy-making authority is shared with (or given to) business, environmental and other organized interests, and even to the increasing use of market-oriented policy instruments (e.g., independent forest certification) to address matters of concern to global civil society. In this process, non-governmental organizations and international institutions have often acted "to reverse the downward effects of globalization on environmental, social and labour standards" (Vogel, 1995). The results of this observable shift on the policy-instrumental axis from public and state-led regulatory measures to private, voluntary and market-driven ones are ambiguous (Pattberg 2005, 2008). While in some cases, private regulations are reported as being more flexible and effective, in others they are considered as being insufficient and remaining non-binding. Notwithstanding this ambiguity, certification initiatives have grown and developed over time, probably becoming the most common and widespread tool for the adoption of corporate social responsibility (CSR) policies and strategies by both public and private, for profit and not-for-profit entities. In this framework, the forest sector represents an interesting – and, from many perspectives, unique – case because, given the multifaceted concept of SFM and complexity of forest-based socio-ecological systems, many standards have been developed.

## 4.2 General concepts of certification: definitions, actors and approaches

Certification can be defined as an attestation (i.e., issuing of a statement) that specifies requirements to be fulfilled related to products, processes, systems or persons (ISO/IEC, 2004). Requirements are normally laid down in the form of standards, i.e., normative documents organized according to a hierarchical and logical structure into principles, criteria and indicators (P, C & I),[1] or at least some combinations of these three hierarchical levels (Lammerts Van Bueren and Blom, 1997). These three are binding elements that constitute the standard itself, i.e., when an organization decides to get certified according to a certain standard, the P, C & I that form the standard become compulsory. In other words, the organization must comply with them in order to obtain certification. In some cases, standards include also verifiers,[2] which are not binding elements but might be useful for auditing (i.e., verification of standard implementation) and management reasons (i.e., correctly interpreting and implementing the standard) (Lammerts Van Bueren and Blom, 1997). Standards are developed and approved by specific standard-setting bodies. Not all existing standards for SFM are designed and can be used for certification purposes (Holvoet and Muys, 2004).

Even though sometimes inappropriately used, the term "certification" strictly refers to a well-defined condition that can be explained through a specific terminology (Box 4.1), mostly adapted from ISO/IEC: Guide 2 (2004). Main actors of any certification process are (i) standardization bodies, that set standards; (ii) organizations, that seeks to be certified; (iii) certification bodies, that check and monitor organizations' compliance with certification rules/standards; and, when applicable, (iv) accreditation bodies, that check and monitor the certification body's compliance with accreditation rules/standards (Bass, 2003).

### Box 4.1 Certification: key terminology

**Accreditation body**: an independent body that provides oversight to certification bodies by assessing them against international voluntary standards (e.g., ISO/IEC 17021).

**Audit**: a systematic exam and evidence gathering process to determine if and to what extent a given organization, products or processes comply with requirements laid down by one or more specific normative documents (e.g., standards).

**Certification body**: an independent body providing auditing and certification/registration services that specify requirements relating to a product, process, system, person or body – as indicated by one or more specific normative documents – are fulfilled.

**Organization**: a single entity or group that achieves its objectives by using its own functions, responsibilities, authorities and relationships. It can be a company, corporation, enterprise, firm, partnership, charity, association or institution and can be either incorporated or unincorporated and be either privately or publicly owned. It can also be an operating unit that is part of a larger entity.

**Standard**: a document, established by consensus and approved by a recognized body, that provides, for common and repeated use, rules, guidelines or characteristics for activities or their results, aimed at the achievement of the optimum degree of order in a given context.

**Standard-setting body**: a body that has recognized activities in standardization, i.e., formulates, issues, updates and implements standards (or other normative documents).

In literature, distinction is normally made among first-, second- and third-party certification.[3] **First-party certification**, also referred to as "self-declaration", refers to the case when an organization states it (or a product/service it produces/delivers) complies with a certain standard, without any assessment of the validity of its statement performed by external bodies. In **second-party certification**, an organization is assessed by a certification body against a certain standard and a statement of conformity is issued to declare the organization itself (or a product/service it produces/delivers) conforms to the standard. This type is also referred to as "declaration of conformity". The certification body, however, is not fully independent from the assessed organization or the standard (e.g., it has some interests in the assessment results, as in the case of a large retailer that audits its suppliers in order to check whether they are respecting or not their commitments). **Third-party certification** consists of an assessment performed by a fully independent certification body, to confirm whether an organization (or a product/service it produces/delivers) complies with a certain (pre-defined) standard or not. Depending on certification schemes and procedures, the certification body might be accredited, i.e., entitled to deliver certificates by an independent accreditation body (Figure 4.1); this is also referred to as "third-party independent and accredited certification", and it is considered the most credible type of certification. The presence of an accreditation body assures that also the certification body is monitored, to guarantee

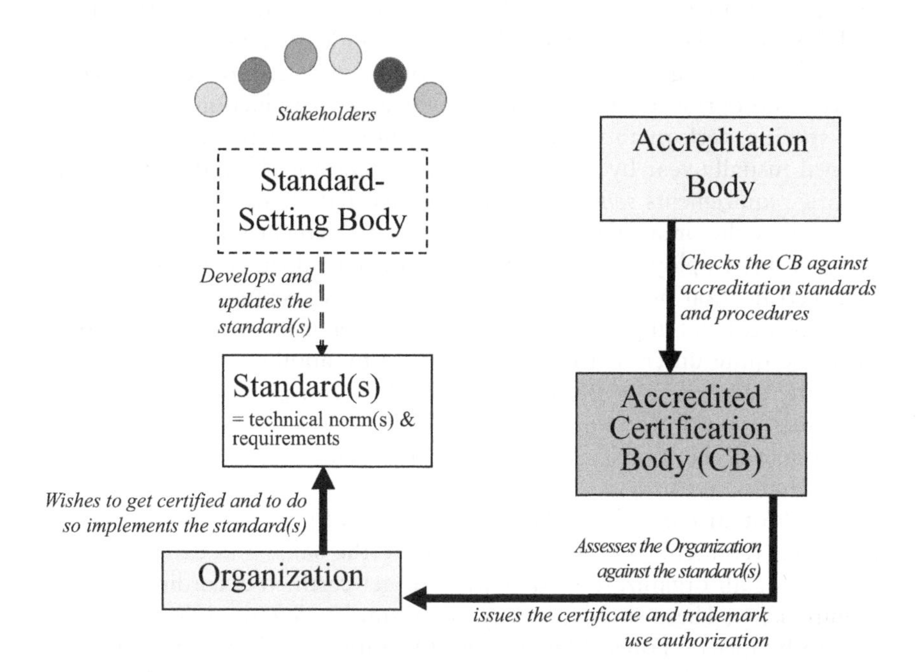


*Figure 4.1* Third-party certification general scheme

Source: Own elaboration

its independency, impartiality and competence in delivering certifications. While, generally speaking, certification and accreditation processes are similar for all certification schemes (e.g., they are all based on audits, they all are voluntary-based, they all require the decision of an organization to obtain a certificate to contract an accredited (or not) certification body for assessing its compliance with the requirements it has committed to), standards might be very different in terms of scope, field of application, approach, etc.

Certification standards are often distinguished into system-based or performance-based standards[4] (Figure 4.2), on the basis of the type of indicators they are made up of. A **system-based standard** builds on the concept of continuous improvement, applied to an organization's policies, management systems and processes (Bass, 1997). This type of standard recognizes and rewards continual progress (improvement) towards achieving performance targets that are not independently and externally pre-established and/or common to all organizations that want to get certified according to the standard itself. In some way, the organization tailors the system to its own objectives, starting points and capacities, i.e., to its specific situation. These standards focus on processes rather than on impacts of operational management practices. To comply with a system-based certification standard, an organization must demonstrate that it has a management system in place and that is regularly kept updated, in order to identify, measure and monitor its impacts (e.g. on the environment or on socio-economic dimensions) as well as progress towards general objectives (e.g., reduction of pollution). But specific performance targets for reaching this general objective are defined (usually year by year) by the organization itself, and there are not specific requirements *sensu stricto* to be respected (apart from those establishing that the organization must have a management plan, or a training program for employees, or an updated storage of documents, etc.). In this case, certification proves that the organization is keeping its management under control and improving it in some way, even if its performance targets might be quite different from those defined by another organization that is certified according to the same scheme. As a consequence, system-based standards, typically, do not allow for labelling with specific trademarks or on-product declarations; rather, it is the management process of the organization that is certified.

A **performance-based standard** outlines a set of performance targets, i.e., minimum requirements for itself, which are the same and externally pre-defined for all organizations that want to get certified according to those requirements. These standards may, for instance, define minimum thresholds to be met by Forest Management Operations and their impacts (e.g., minimum set aside area within the forests, maximum extension of clearcutting, etc.) or acceptable/unacceptable actions and behaviours (e.g., the use of GMOs or of certain chemicals/pesticides). Performance targets can be less or more strict/high, thus being lower or higher demanding for the organization that is seeking to be certified with respect to its own initial capacity, management practices, etc. (Bass, 1997). In this case, certification proves that all certified organizations have been able, at a minimum, to reach certain performances in their operations. As a consequence, performance-based certification standards typically allow the use of specific trademarks and on-product declarations on certified products.

While a system-based standard is mainly characterized by *descriptive* indicators (e.g., reforested area within a certain period, yearly harvesting rates, percentage of forest plantations over total forest area, etc.), aiming to organize the management system for continuous improvement and assess progresses, a performance-based standard is mainly characterized by *prescriptive* indicators that facilitate organizing the management system in order to fully comply with the pre-defined minimum performance levels. These two kinds of indicators do not exclude each other. Rather, they coexist and complement,[5] so that standards do normally include a combination of the two.

Certification schemes differ also in terms of accreditation system, internal governance mechanisms, certification procedures and practices (e.g., sampling strategies used in audits) and labelling rules (e.g., use of claim and logo) (Nussbaum et al., 2002).

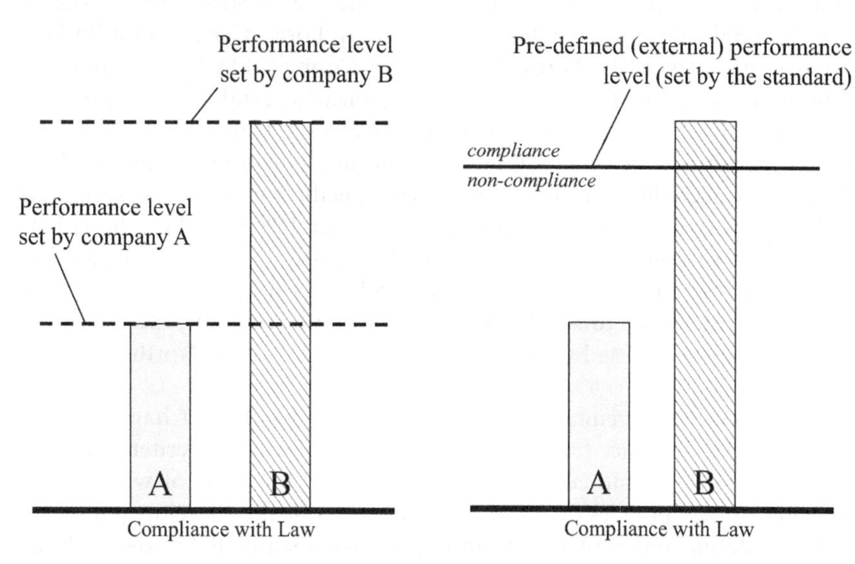


*Figure 4.2* System-based and performance-based approaches

Source: Own elaboration

## 4.3 Overview of certification schemes that might operate in the forest sector

Forest certification refers specifically to those certification initiatives focused on and dedicated to promoting sustainability of forest management practices and to increasing the value of forest products on the market through the use of labels. However, many other certification schemes can be adopted in forestry, such as cross-sectoral schemes (e.g., quality or Environmental Management Systems) or very specific schemes covering special issues linked with forest management (e.g., carbon). This paragraph provides a short overview of (a) certification schemes specifically intended for forestry, and (b) certification schemes that might be used also in forestry without being designed for this field of application, or that address very specific forest-related issues.

### *4.3.1 Forest-specific certification schemes*

These initiatives started in the early 1990s, just after the Rio de Janeiro Conference. They include standards for certification of both forest management (i.e., management practices at the Forest Management Unit level) and

Chain of Custody (CoC) (i.e., product traceability), thus at the end allowing labelling of forest products coming from certified forests. At global scale, two schemes dominate forest certification: the Forest Stewardship Council (FSC®) and the Program for the Endorsement of Forest Certification (PEFC).

Created in 1993, the **Forest Stewardship Council (FSC)**[6] is an independent, non-governmental, not-for-profit organization established to promote the responsible management of the world's forests. It was the first organization worldwide creating, establishing and implementing an independent, third-party accredited certification system specific for the forest sector. As of January 2018, an overall forest area of about 198.86 million ha is FSC certified in 83 countries around the world, mostly (84%) concentrated in Europe (49%) and Northern America (35%). As for FSC Chain of Custody certificates, they total 33,626 units worldwide, covering 122 countries, with a prevalence being European (53%), Asian (31%) and North American (10%).

For Forest Management Certification (more details in Chapter 5), its globally valid reference framework are 10 principles and 56 criteria (P&C)[7] for Forest Stewardship consolidated in 2002, on the basis of which additional rules (e.g. policies or standards) have been further developed in order to better define and explain certain requirements stipulated in the 10 basic principles. International FSC P&C represent the common starting point and "benchmark" for the definition of local/national FSC standards, which have to be developed according to FSC specific procedures and approved by FSC itself. FSC has recently reviewed its P&C, publishing a new version (5–2) in 2015, and developed a set of International Generic Indicators (IGIs) to facilitate the development of local/national FSC standards. However, in those areas where any national or local Forest Stewardship Standard has not yet been approved, certification bodies may carry out certification according to "generic" (or *ad interim*) standards, which are developed by the accredited certification bodies themselves consistently with FSC P&C and adapted – with input from local stakeholders – to account for the local conditions in the country or region in which they are to be used.

FSC P&C apply both to natural forests and forest plantations. For the latter, a dedicated Principle (the last one, Principle 10) was adopted in 1998 in addition to existing ones. However, version 5–0 of FSC P&C no longer presents such a specific Principle dedicated to plantations. Rather, plantation-specific requirements are covered throughout all P&C.

FSC Forest Management Certification allows for both individual or group certification. **Individual certification** is based on one forest owner (or forest manager, or any other type of entity, i.e., the organization in charge of forest management) directly applying for certification as a single entity, in relation to a single property (that, however, can be geographically fragmented). **Group certification** is based on a group entity applying for certification, where group members are more than two forest owners, forest managers and/or any other type of organization in charge of forest management, including their associations and/or combinations of them, all committed to the same FSC relevant standards and acting as a group on the basis of agreed rules and procedures (group membership requirements and regulations). The group entity holds the Forest Management Certificate (not the single group members), having the responsibility towards the certification body for guaranteeing that FSC standards requirements are fully met in all forest properties participating in the group. As long as group members comply with group membership requirements, their forest ownerships are covered by the certificate issued to the group entity. Special sampling strategies are defined for auditing a group.

This type of certification has been designed in order to facilitate and thus increase the access of small forest owners to Forest Management Certification. With the same purpose, specific and simplified certification procedures have been defined for the so-called Small and Low Intensity Managed Forests (SLIMF).[8] FSC also has a number of standards and specific procedures for regulating Chain of Custody certification (more details in Chapter 6), which apply to wood processing industries and traders.

Created in 1998, under the initial name of Pan-European Forest Certification and as an alternative to FSC certification for European forests, the **Program for the Endorsement of Forest Certification (PEFC)** assumed the current denomination in 2004, when it enlarged its area of activity to include also non-European countries and their certification schemes, starting from the USA and Canada. PEFC is an umbrella organization that endorses national forest certification systems; national certification systems that have developed standards in line with PEFC requirements can apply for endorsement to gain access to global recognition by the PEFC Council.[9] To achieve endorsement, they need to meet PEFC requirements and undergo specific assessment procedures. As of December 2017, 39 national schemes have been endorsed; an overall forest area of about 313 million hectares is PEFC certified in 37 countries around the world, mostly (86%) concentrated in Europe (31%) and North America (55%). As for PEFC Chain of Custody certificates, they total 11,484 units worldwide, covering 72 countries, mostly concentrated in Europe (82%).

Forest management certification is released in conformity with PEFC-endorsed national standards that must be coherent with basic requirements defined by the PEFC ST 1003:2010 standard. In particular, PEFC national standards shall respect the results of regional intergovernmental processes for the definition of sustainable forest management principles and criteria. For temperate forests, reference shall be made to the common framework of Pan-European Criteria for Sustainable Forest Management and the related Pan-European Operational Level Guidelines (PEOLG), that provides interpretation of the framework criteria for forest management practices. For tropical forests, the reference frameworks are the ATO/ITTO Principles,

Criteria and Indicators for the sustainable forest management of African natural tropical forests (at least for those countries this document refers to). For all other countries, criteria and indicators proceeding from the relevant regional intergovernmental processes (e.g., Tarapoto process for Amazon forests or Lepaterique Process for Central American ones) should establish the basis for the development of national certification standards looking for PEFC endorsement. Requirements laid down by PEFC ST 1003:2010[10] are applicable to all types of forests and their interpretation according to different forest types or geographical zones is provided as an appendix. In a similar way, specific interpretation is provided for forest plantations.

Also, PEFC reckons **individual** or **group Forest Management Certification**, the latter being regulated by a specific standard.[11] However, it has to be recalled that, until October 2010, a third alternative form of Forest Management Certification was allowed under PEFC rules, the so-called **regional certification**. This type of Forest Management Certification was specific to PEFC only. Regional certification was a multisite certification of forests within delimited geographic boundaries. In this case, the applicant for certification had to represent forest owners/managers representing or managing more than 50% of the forest area in the region. Only the forests of participating forest owners/managers were considered as certified. Their area counted as certified area and the forest raw material coming from there was considered certified raw material. Several regional certificates were created in the 12 years between the launch of PEFC in 1998 and the reform introduced in 2010, which now have been transformed into group certificates.

While international schemes, such as FSC, or mutually endorsed national schemes, such as PEFC, cover most of certified forests worldwide, some national forest certification schemes remain outside these frameworks. This is the case, for example, of the **Lembaga Ekolabel Indonesia (LEI)**,[12] i.e., the Indonesian Ecolabelling Institute, a nonprofit organization that has developed forest certification standards specific to the Indonesian forests. The LEI system includes four different schemes, i.e., three sets of standards for forest management – for natural forests, plantations and community forests – and one for Chain of Custody certification. According to the most recently available figures, certified forests according to LEI standards cover a total area of about 1.87 Mha, 78% of which is represented by plantations, while six CoC certificates have been issued against LEI standards. LEI and FSC recently signed a Memorandum of Understanding.

Aside from the previously mentioned forest certification schemes, specifically for CoC certification, it is important to note that the **International Organization for Standardization (ISO)** has established a Project Committee (ISO/PC 287) to develop an international standard (ISO/NP 38001) defining CoC requirements for wood, wood-based products and lignified materials other than wood (such as bamboo and cork). The standard[13] is still under development and aims to unify current standards on traceability of wood forest products.

### *4.3.2 Certification schemes that might be used in forestry or address very specific forest-related issues*

Although not specifically developed for the forest sector, many certification standards can be used to issue certificates covering forest management or forestry-related (e.g., wood or paper processing) operations and products. These include, for example, **ISO standards** for Quality Management Systems, QMS (ISO 9001) and Environmental Management Systems, EMS (ISO 14001). These standards are characterized by a system-based approach, and promote no substantive standard to benchmark performance against other operators. It is an incremental approach to achieving sustainable forest management that relies on the concept of continuous improvement, while remaining vague with regard to its extent, timeliness and coverage (Cashore et al., 2006). In 1998, ISO developed a specific Technical Report (ISO/TR 14061:1998) to assist forestry organizations in the use of EMS standards ISO 14001 (and ISO 14004); the document, however, has been withdrawn.

Other system-based standards that are not forest-specific but also address issues of interest for the forest sector are the internationally applied **British Standard for Occupational Health and Safety Management Systems** (BS OHSAS 18001), which is largely applied in alignment with ISO 9001 and ISO 14001 management systems and allows organizations to identify and control health and safety risks, reduce the potential for accidents and thus improve their overall health and safety performances. However, it has to be mentioned that workers' health and safety requirements are normally included in Forest Management Standards (e.g., FSC and PEFC ones) and have been recently added also to CoC standards.

In addition, social accountability certification standards, with a similar system-based approach, can be implemented in forestry. Among them, a relevant role is played by **SA8000**, developed by Social Accountability International (SAI)[14] as one of the world's first social certification standards that operates across all industrial sectors. It is based on the UN Declaration of Human Rights, conventions of the International Labour Organization (ILO), as well as relevant national laws. Social accountability requirements spread over nine specific areas: child labour, forced or compulsory labour, health and safety, freedom of association and right to collective bargaining, discrimination, disciplinary practices, working hours, remuneration and the overall management system. Some of these social accountability requirements might be identified as being already included in Forest Management Standards (e.g., where they ask for regular working conditions and respect of labour rights of forest workers).

Given the importance of small-scale and community forestry, there has been a growing interest in developing specific certification solutions for smallholders. Some of these initiatives were directly developed by forest certification schemes as part of their standard-setting activities – e.g., the development of group certification or SLIMF requirements and procedures – and some others were launched in cooperation with other organizations, such as, for example, Fair Trade standards. This is the case of a specific dual certification project created by FSC and the **Fairtrade Labelling Organization (FLO)**.[15] Further details are provided in Chapter 10.

A special case of certification that might be applied to forest products is represented by standards declining their prevalent performance-based approach into a **Life-Cycle-Assessment (LCA) methodology**. This means that all potential environmental impacts (raw materials, energy and water consumption, emissions in the atmosphere, etc.) from the production and trade of a certain product are taken into consideration from the very beginning (i.e., material sourcing and product designing) to the very end (dismissing after end use). An example is the **EU Ecolabel**, established according to Regulation (EC) 66/2010. Building on an LCA approach, specific requirements are developed for single product categories. At the moment, specific criteria have been defined for 34 product types under 13 different product categories, including paper products (converted, newsprint, printed, copying and graphic and tissue paper), wooden floor coverings and wooden furniture for which requirements regarding the sourcing of forest materials have been defined.[16] A similar approach is followed by many nationally or regionally defined standards such as the Blau Angel[17] (Germany) and the Nordic Ecolabel, or Nordic Swan[18] (Scandinavian countries). Since these standards are intended for assessing product-specific performances and assuring minimum performance levels are reached, they all allow labelling compliant products with specific on-product trademarks and claims.

Forest certification schemes are exploring and testing the possibility to include management and provisioning of ecosystem services within their standards and procedures, like in the case of version 5–0 of FSC P&C or the pilot project ForCES (Forest Certification of Ecosystem Services), launched by FSC.[19] Forest management requirements can already be adopted for the purpose of **carbon projects certification** with reference to initiatives of afforestation and reforestation, Improved Forest Management, as well as Reducing Emissions from Deforestation and Forest Degradation (REDD) projects.

An increasing number of sustainability standards are also developed for the growing wood energy sector. An example is the **Sustainable Biomass Program (SBP)**, created in 2013[20] by the major European utilities (including Dong Energy, Drax, E.ON, SDF Suez, RWE and Vattenfall) using biomass, mostly wood, in the form of pellet, for heating purposes. In addition to initiatives related to biomass sustainability, biomass quality standards like **ENPlus** are becoming very appreciated by the consumers.

Apart from wood and wood-based products, certification can also apply to collection, production and processing of **Non-Wood Forest Products (NWFPs)**. NWFPs are covered by forest management as well as CoC standards and in some cases *ad hoc* local Forest Management Standards have been developed, such as, for example, FSC standards for Brazilian nuts in Bolivia and Peru, and bamboo in Colombia. However, other standards specifically intended for, or potentially applicable to, NWFPs can be mentioned. For example, the **FairWild Foundation**[21] has developed and made available a standard and certification scheme covering ecological and social aspects dealing with the collection of products from the wild (e.g., medicinal and aromatic plants, berries, wild fruits, nuts and seeds, mushrooms). A number of NWFPs are registered (or have applied for registration) for the EU schemes and policies that promote and protect **geographical indications** and **traditional specialities** for agricultural products and foodstuffs (EU Regulation No 1151/2012 and related norms). For NWFPs, **organic farming/production certification standards** can be used as well, such as for example standards developed by the **International Federation of Organic Agriculture Movements (IFOAM)**[22] or, in the case of EU producers, standards developed in compliance with Regulation (EC) 834/2007.[23]

## 4.4 Final considerations

Currently, one of the main drivers of forest certification remains the demand for certified forest products on the European and North American markets. Despite unclear or contradictory information on forest certification, real impacts on the economic, social and environmental performance of certified organizations, both Chain of Custody and Forest Management Certifications have been growing exponentially during the last 20 years worldwide (and they are still growing). This might be due to the high capacity of some advocacy coalitions (e.g., those based on environmental and social NGOs) to increase awareness of, and thus demand for, forest certification, mainly based on labels that are becoming quite widespread, if not yet famous, at least in Western countries (i.e., FSC and PEFC). In the future, with the economic growth and demographic increase of population in emerging countries, both the environmental awareness and market demand for forest products (i.e., paper, industrial timber for building, wood biomasses for energy production) are expected to rise even more. Therefore, it is likely a further development of forest certification, especially in emerging and exporting countries. However, over 20 years after the first forest obtained certification, only 11% of forests worldwide are certified (and no data are available on the overall volume of certified timber produced and traded) (UNECE/FAO, 2016). This result, so different with respect to the stated goals when the concept of forest certification was launched, is most likely associated with the various and challenging constraints to business environment and institutional context determined by poor governance conditions (e.g., corruption, illegalities, weak legal frameworks, unclear or unallocated tenure rights) in many exporting developing countries. Lack of technical skills to operationalize forest certification with local experts within a country, and the consequential needs (and costs) of contracting foreign certification bodies (for auditing and issuing certificates) and/or foreign consultants (for technical assistance in preparing management systems of forest organizations to comply with requirements and thus obtain certification) have been other obstacles to certification uptake in the past.

Currently, knowledge about mechanisms, practices and procedures of forest certification are consolidated among certified organizations, consultants, certification bodies and accreditation entities, including, for example, standard-setting, external auditing, internal monitoring, periodic surveillances, corrective action requests, individual or group certification rules, etc. A lot of information is now available on websites and in the documents of standardization, certification and accreditation bodies.[24] International, national and sometimes even local NGOs make online reports, data and information on forest certification available. Due to the synergic activities of certified companies, certification programs, NGOs and governments, the standards and logos of the two most important forest certification schemes, and indeed, the other environmental and social responsible standards that can be used also in forestry (e.g., Fair Trade, organic farming, carbon sequestration), are nowadays well known. The international debate, and sometimes harsh conflicts, between supporters of NGO-led and those of industry-led certification programs (Cashore et al., 2006) have brought positive effects to some, determining some convergences between PEFC and FSC standards (e.g., getting more strict performance requirements[25]).

Forest certification, not only as a market-based instrument (Cashore et al., 2003) but also as a predominant worldwide "discourse, based on ideas, beliefs and assertions", has been able to induce policy changes (Humphreys, 2009) and changes of policy impacts on forest management. However, empirical evidence of forest certification impacts based on robust scientific-based methodology (able to provide an independent wide evaluation) are still very limited. Further research is required to cover this knowledge gap.

## Notes

1 Principles are fundamental rules or guiding ideas for action, i.e., explicit elements of a goal in relation to relevant aspects of forest-based socio-ecological systems (SES) (e.g., well-managed forests, protected forest-based community rights). Criteria are states or aspects of forest-based SES that should be in place as a result of adherence to a principle, able to provide more specific insights with respect to principles about relevant issues for management. Indicators are quantitative or qualitative parameters that can be assessed in relation to criteria; they describe objectively verifiable features of the SES (Lammerts van Bueren and Blom, 1997).

2 Verifiers are possible sources of information for the indicators or reference values for indicators (Lammerts van Bueren and Blom, 1997).
3 First-, second- and third-party audits are also reported in literature, but – as clarified in Box 3.1 – audits should not be confused with certification.
4 Because these two types are often complementary (i.e., most of the certification standards include both types of indicators), it is more correct to refer them to as "prevailing" system-based standards or prevailing performance-based standards.
5 See note 3.
6 More details in Chapters 5 and 6. Official documents and updated statistics at: www.fsc.org.
7 FSC Principles and Criteria for Forest Stewardship, FSC-STD-01–001 (version 4–0). Forest Stewardship Council, Bonn.
8 In the SLIMF definition, "small" means forests smaller than 100 ha and "low intensity" stands for forests where the rate of harvesting is less than 20% of the Mean Annual Increment (MAI) and either the annual harvest from the total production forest area is less than 5,000 m3 or the average annual harvest from the total production forest is less than 5,000 m3/yr during the period of validity of the certificate. FMUs consisting of natural forests in which only Non-Timber Forest Products are harvested are considered as SLIMF.
9 Global statistics on endorsed PEFC certifications schemes: www.pefc.org.
10 PEFC International Standard – PEFC ST 1003. (2010). *Requirements for Certification schemes: Sustainable Forest Management: Requirements.* PEFC Council, Geneva.
11 PEFC International Standard – PEFC ST 1002. (2010). *Requirements for Certification Schemes: Group Forest Management Certification: Requirements.* PEFC Council, Geneva.
12 Updated figures on LEI certified forests and many other details can be found at: https://lei.or.id/.
13 ISO/PC 287 Chain of custody of wood and wood-based products. www.iso.org/iso/iso_technical_committee?commid=4952370.
14 SAI (2014). *Social Accountability 8000 International Standard. SA8000: 2014.* Social Accountability International, New York.
15 FLO (2010). *Fairtrade Standards for Timber for Forest Enterprises Sourcing from Small-Scale/Community-Based Producers.* Fairtrade Labelling Organization, Bonn.
16 EU Ecolabel product groups and criteria are available at: www.ec.europa.eu/environment/ecolabel/products-groups-and-criteria.html.
17 See also: https://www.blauer-engel.de. Among covered product groups: particle- and fibreboards, paper, paperboard, plywood and wood pellets.
18 See also: www.svanen.nu. Among covered product groups: biofuel pellets, durable wood products, copy and writing paper, tissue paper, indoor and outdoor furniture and windows. Nordic Ecolabel. URL: www.nordic-ecolabel.org/
19 See: www.forces.fsc.org/
20 Initially created as the Sustainable Biomass Partnership.
21 FairWild (2013). *Guidance for Industry.* FairWild Foundation, Weinfelden.
22 IFOAM (2014). The IFOAM Norms for Organic Production and Processing. Version 2014.
23 Council Regulation (EC) No 834/2007 of 28 June 2007 on organic production and labelling of organic products and repealing Regulation (EEC) No 2092/91. Official Journal of the European Union, L 189/1, 20.07.2007. An indicative list of national or specific organic standards for collection of wild plants developed according to the EC Regulation is available at http://organicrules.org/custom/differences.php?id=2aaf.

24 It is worthwhile to mention that FSC, SAI, FLO and IFOAM are all members of the International Social and Environmental Accreditation and Labelling Alliance (ISEAL), an association of leading voluntary international standard-setting and conformity assessment organizations that focuses on social and environmental issues. For further information: www.isealalliance.org.
25 After the 2010 PEFC General Assembly, clear requirements have been introduced by PEFC for example about the prohibition of Genetically Modified Organisms in PEFC national standards or conversion of natural forests into plantations.

## References

Bass, S. (1997). Introducing Forest Certification. A report prepared by the Forest Certification Advisory Group (FCAG) for DGVIII of the European Commission. European Forest Institute, Discussion Paper 1, Joensuu, Finland.

Bass, S. (2003). Certification in the Forest Political Landscape. In: Meidinger, E., Elliott, C., Oesten, G. (eds.). *Social and Political Dimensions of Forest Certification*. Kessel Publishing House, 27–49. Available at www.forsthbuch.de

Cashore, B., Auld, G., Newsom, D. (2003). Forest Certification (Eco-Labelling) Programs and Their Policy-Making Authority: Explaining Divergence among North American and European Case Studies. *Forest Policy and Economics 5*: 225–247.

Cashore, B., Gale, F., Meidinger, E., Newsom, D. (eds.) (2006). *Confronting Sustainability: Forest Certification in Developing and Transitioning Countries*. Yale School of Forestry and Environmental Studies Publication Series Report Number 8. Yale SF&ES, New Haven.

Holvoet, B., Muys, B. (2004). Sustainable Forest Management Worldwide: A Comparative Assessment of Standards. *International Forestry Review* 6(2): 99–124.

Humphreys, D. (2009). Discourse as Ideology: Neoliberalism and the Limits of International Forest Policy. *Forest Policy and Economics 11*(5–6), October 2009: 319–325, ISSN 1389–9341.

ISO/IEC (2004). *ISO/IEC Guide 2:2004: Standardization and Related Activities: General Vocabulary*. International Organisation for Standardisation, Geneva.

Lammerts Van Bueren, E.M., Blom, E.M. (1997). *Hierarchical Framework for the Formulation of Sustainable Forest Management Standards (Principles, Criteria and Indicators)*. The Tropenbos Foundation, Leiden.

Lemos, M.C., Agrawal, A. (2006). Environmental Governance. *Annual Review of Environmental and Resources 31*: 297–325.

Nussbaum, R., Jennings, S., Garforth, M. (2002). Assessing Forest Certification Schemes: A Practical Guide. Proforest, UK. Available at www.proforest.net//proforest/en/files/assessing-schemes.pdf

Pattberg, P. (2005). What Role for Private Rule-Making in Global Environmental Governance? Analyzing the Forest Stewardship Council (FSC). *International Environmental Agreements: Politics, Law and Economics 5*(2): 175–189.

Pattberg, P. (2008). Private Governance and the South: Lessons from Global Forest Politics. In: Mitchell, R.B. (ed.). *International Environmental Politics*. Sage, London.

UNECE/FAO (2016). *Forest Products Annual Market Review 2015–2016: Forestry and Timber Section*. United Nations, Geneva.

Vogel, D. (1995). *Trading Up: Consumer and Environmental Regulation in a Global Economy*. Harvard University Press, Cambridge.

Chapter 5

# Certified forest products and services

## A market snapshot

*Lucio Brotto, Diego Florian, Enrico Vidale, Alex Pra, Nicola Andrighetto, Ariadna Chavarria, Giulia Corradini and Alessandro Leonardi*

This chapter gives a general overview of the market for certified forest products and services, providing references to up-to-date studies and market analysis for further information. The chapter is divided into five paragraphs corresponding to the market analysis for FSC-certified products (paragraph 5.1), forest carbon finance (paragraph 5.2), wood energy (paragraph 5.3), Non-Wood Forest Products (paragraph 5.4) and other ecosystem services (paragraph 5.5).

### 5.1 Market for FSC-certified products

#### *5.1.1 Global overview*

The Forest Stewardship Council (FSC®) certification system is recognized in about 80 countries globally, from Forest Management (FM) Certification through to Chain of Custody (CoC) certification. FSC maintains its system credibility. The FSC system is one of the earliest examples of performance-based standard processes, grounded on environmentally appropriate, socially beneficial and economically viable FM practices around the world. As of January 2018, an overall forest area of about 198.86 million ha is FSC certified in 83 countries around the world, with 33,626 CoC certificate units covering 122 countries. According to FSC (2017), FM Certificates rose by 35% over the last 5 years, and the FSC-certified forest area by 32% (Table 5.1). The FSC-certified forest area is mostly concentrated in Europe and North America, in particular, Canada and Russia, as detailed in Table 5.2. The number of FSC CoC certificates increased by 42% over the last 5 years; the countries with the highest numbers of certificates are presented in Table 5.3. FSC is represented in 42 countries by Network Partners that are working to promote responsible forest management and to bring FSC-certified products and materials from forests to stores, together with non-governmental

*Table 5.1* Evolution of FSC FM Certificates, certified forest area and CoC certificates

| *Years* | *FSC Forest Management Certificates* | *FSC-certified forest area (million ha)* | *FSC CoC certificates* |
|---|---|---|---|
| 2012 | 1,084 | 149 | 22,230 |
| 2013 | 1,175 | 172 | 24,789 |
| 2014 | 1,265 | 182 | 27,316 |
| 2015 | 1,311 | 185 | 28,604 |
| 2016 | 1,267 | 184 | 29,801 |
| 2017 | 1,462 | 196 | 31,599 |

Source: Adapted from FSC (2017)

*Table 5.2* Countries with largest FSC-certified forest area

| *Country* | *FSC-certified forest area (million ha)* | *Percentage of total forest cover* |
|---|---|---|
| Canada | 54.6 | 16 |
| Russia | 43.7 | 5 |
| United States | 13.7 | 4 |
| Sweden | 12.2 | 44 |
| Belarus | 8.4 | 98 |
| Poland | 6.9 | 74 |
| Brazil | 6.2 | 1 |
| Ukraine | 2.8 | 30 |
| Indonesia | 2.8 | 12 |
| Congo, Republic of | 2.6 | 13 |
| Chile | 2.3 | 9 |
| Gabon | 2 | 52 |
| United Kingdom | 1.6 | 61 |
| South Africa | 1.3 | 6 |
| Estonia | 1.3 | 13 |
| Finland | 1.3 | 1 |
| New Zealand | 1.2 | 10 |
| Australia | 1.2 | 1 |
| Germany | 1.1 | 10 |
| Lithuania | 1 | 50 |

Source: Adapted from FSC (2017)

organizations (NGOs), companies and private citizens representing FSC members. The FSC certification scheme is based on the promotion of a forest production model that improves, at the same time, the management of forestry practices and the local communities' benefits, while receiving economically viable services.

### 5.1.2 *Supply and demand of FSC-certified products*

According to FSC (2016), the majority of FSC-certified businesses are paper-based products, particularly printed materials, traded by companies

*Table 5.3* Countries with the highest numbers of FSC CoC certificates

| *Country* | *Number of CoC certificates* |
|---|---|
| China | 4,841 |
| United States | 2,746 |
| United Kingdom | 2,364 |
| Germany | 2,202 |
| Italy | 2,067 |
| Poland | 1,466 |
| Netherlands | 1,235 |
| Japan | 1,115 |
| Brazil | 1,083 |
| Spain | 844 |
| France | 743 |
| Canada | 734 |
| Hong Kong | 617 |
| Vietnam | 532 |
| Switzerland | 474 |
| Russia | 428 |
| India | 354 |
| Sweden | 340 |
| Latvia | 306 |
| Belgium | 288 |

Source: Adapted from FSC (2017)

in Europe and Asia. The share of FSC-certified timber and sawn wood sales increased in recent years, both in Europe and North America, mainly thanks to the expansion of the green building sector. Schemes like the USA's Leadership in Energy and Environmental Design (LEED),[1] the International Green Construction Code (IGCC) and the Building Research Establishment's Environmental Assessment Method (BREEAM)[2] use FSC certification schemes to evaluate responsible forestry operations. LEED and BREEAM, for example, provide credits when FSC-Certified Wood products are used in the buildings. As a result of the high demand for FSC fibreboards, FSC is promoting, through a campaign, the commitment of potential FSC-certified fibreboard buyers as a way to create incentives for fibreboard producers.[3] Other FSC-certified products that are growing in terms of market share are labels and hang tags, toys and crafts, fibres and building-related products such as wooden doors and windows frames.

FSC-certified products can be divided into three categories:

- FSC 100% sources, based on 100% FSC-certified material. This is made mostly of solid wood products and construction wood.
- FSC Mix sources, based on certified materials mixed with recycled or Controlled Wood. This is made mostly of wood particles and paper-based products.
- FSC Recycled.

The supply of FSC-certified products is growing rapidly in Latin America and Europe (FSC, 2016). Market demand remains the strongest driver for sourcing FSC-certified products as many companies plan their sourcing in relation to market demand. In this sense, FSC can help businesses and producers to be less dependent on volatile market developments, as sourcing of certified materials can be better planned in advance. The demand for FSC-certified materials is increasing: the number of FSC Certificates has been growing significantly, stimulated also by the FSC efforts to increase the market visibility. The demand is mostly increasing for certified products like tissue paper, Non-Timber Forest Products (e.g., rubber, food, drinks, cosmetics and bamboo) and pulp and packaging. Globally, the demand for FSC products is increasing both in Northern and Southern Hemisphere countries. Russia, India, Brazil and China represent 30% of the global demand for FSC-certified products and their share is likely to increase as their economies develop further (FSC, 2016). Detailed data on FSC-certified products market can be found in the last FSC Global Market Survey (FSC, 2016).

### 5.1.3 Reasons for certification

Reasons for obtaining certification differ from user to user, depending on user type, size, location, mission, marketing strategies and many other aspects.

#### 5.1.3.1 Reasons for private companies

The most important category of certification users are business-oriented private companies (e.g., forest managers/producers of wood raw material, semi-finished or finished products and traders). The main reasons why this category seeks certification include:

- To maintain or improve market access/share, to acquire a positive and credible "green" image and reputation.
- To obtain a premium price for certified products and to reduce business risks by securing tenure and concession, increasing price and market stability (Carlsen et al., 2012).
- To strengthen efficiency and capacity and to gain political recognition and influence.
- To ensure good image and reputation, in particular for large-scale timber firms, which are more vulnerable to public scrutiny than small and medium enterprises (SMEs). In this sense, certification is used as a social "license to operate".
- To attract environmentally responsible investors (Takahashi et al., 2003).
- To adequately respond to stakeholder pressure like activism or environmentalism (Bouslah et al., 2010).

However, often, most of the above-mentioned motivations are expectations rather than real consequences and impacts of certification on private companies, for which empirical evidence and studies are still unclear and sometimes contradictory (Romero et al., 2013). For example, evidence of neutral (or even negative) impacts of certification on financial performance of firms have been reported for Canada and the USA (Bouslah et al., 2010). According to this study, differences exist between the short-term and long-term results:

> The results of short-run event returns indicate that forest certification does not have any significant impact on firm financial performance, regardless of the certification system carried out by firms. In contrast, the long-run post-event abnormal returns vary according to certification systems. However, in the long-run, the financial impact of forest certification depends on who grants the certification, since only industry-led certification (SFI, CSA and ISO14001) is penalized by financial markets. The NGO-led FSC certification is not penalized.
>
> (p. 569)

These results suggest that, at least in the long-term, companies might have more or fewer advantages from forest certification depending on the scheme they decide to obtain certification with; "selecting certification recognized and supported by environmental groups" (Bouslah et al., 2010) might contribute to enhancing the long-term reputation of firms, thus improving also financial performance. However, it should be underlined that these conclusions refer to pure financial performances (increasing or not shareholder returns), thus not taking into account the economic value of environmental or social benefits associated with forest certification programs voluntarily adopted by private companies.

Similarly, the existence and amount of a premium price for certified timber with respect to uncertified timber has been strongly debated for years, with limited, fragmented, elusive and/or contradictory empirical evidence; while, for example, a premium price of 27% to 56% is reported for certified logs of high quality hardwoods in Sabah, Malaysia (Kollert and Lagan, 2007), most Japanese companies did not receive any premium price for their certified forest products (Owari and Sawanobori, 2007). Several studies (Romero et al., 2013) based their conclusions on the existence of a premium price ranging from 2% to 30% on the estimation of European and US consumers' willingness to pay (Kollert and Lagan, 2007), which is not necessarily transformed into real buying behaviour.

Aside from uncertainties about direct or indirect economic results, many factors are recognized to affect certification uptake (Carlsen et al., 2012): market conditions, compliance costs, state-led governance, norms/attitudes/

practices, general awareness and organizational capacity of certifying institutions. Especially in developing economies, "the ability of firms to organize themselves and act collectively in their certification efforts and the ability of certification organizations to establish local networks capable of supporting certification efforts at various stages" as well as "government support" are reported as more important factors for certification uptake than market conditions (Carlsen et al., 2012, p. 84). However, also market conditions matter, including competitors' marketing strategies, forest exports and demand for certified products, that are likely to differ depending on the certification standard (e.g., higher for NGO-led certification such as FSC because of the strong support and pressure of environmental movements). Companies will therefore select one or more certification type on the basis of their potential or real market demand, probably giving preference or priority to those standards that allow for labelling of their forest products (e.g., with FSC, PEFC, FLO or other logos, or a combination of them), thus enabling final consumers to identify and buy them in the shops.

Direct and indirect costs of complying with certification standards are reported as relevant factors affecting certification decisions. At least in tropical countries such as Ghana, Bolivia and Ecuador, "high compliance costs discourage [firms'] engagement in certification especially when limited knowledge is provided about certification requirements in general, and about compliance costs specifically" (Carlsen et al., 2012, p. 90). However, it is recognized by practitioners that indirect costs can be high, as they include all those costs related to activities, procedures and document development, and training required to convey managerial and operational practices in line with certification requirements. They might include, for example, a forest inventory as well as the development/implementation of a forest certification plan, if lacking. The absence of an approved forest management plan, or the presence of major gaps in the current management practices with respect to the certification standards requirements can inhibit the decision of applying for certification.

Other factors influencing the application of forest certification are the regulatory environment – the stricter the existing regulatory prescriptions are, the easier the compliance with certification requirements by firms seems to be, thus positively influencing the decision of getting certified (Auld et al., 2008) – and vice versa, lack of effective, transparent and predictable business environments or other weak governance conditions (corruption) are potential obstacles in certification uptake especially for medium and small enterprises (Ebeling and Yasué, 2009). Interestingly, the human and social capital (literacy rate and citizen empowerment, especially of women) seems to positively influence the growth of Forest Management Certification in a country (van Kooten et al., 2005).

#### *5.1.3.2 Reasons for public administration*

Other actors that might be interested in obtaining certification are public forest administrations owning and managing forests and who have, as a main goal, environmental or social protection rather than timber or production of Non-Wood Forest Products (NWFPs). The need to improve the ecological and social conditions of their forests is often an important motivation for public administration, by aligning their ordinary forest management practices to accept and consolidate sustainable forest management requirements (i.e., standards defined by forest certification schemes). Indirect goals for this category of forest certification users might be to protect biodiversity, to reduce deforestation and/or to reduce the negative impacts of forest management activities on soil, air and water (Romero et al., 2013). In this case, labelling of forest products can be useless, and organizations might decide to obtain certification only according to those schemes that do not allow for a product label but assist managers in reducing environmental pollution and/or to improve internal management processes and procedures (e.g., ISO 14001). One more deciding factor for these public organizations (usually linked to politicians) to obtain certification (or not) is to maintain or create a good reputation of public opinion, as citizens and media are able to influence elections.

#### *5.1.3.3 Reasons for NGOs*

Another category of certification users includes NGOs. They can both manage/hold forests for timber production/commercialization and/or for environment/local community protection. While their economic motivations are likely similar to those reported for private companies, their environmental/social motivations are expected to be similar to those reported for public forest owners/managers and not oriented to timber production. However, because of their nature and mission, some NGOs might have special motivations (e.g., local community empowerment, institutional capacity building) connected with small-scale forestry and Fair Trade issues, such as the creation of small landowner associations and networks that have higher contractual power on the market, or the use of certification labels to give the community the pride of being known and recognized at international level. As for the other categories, NGOs' decisions to obtain certification are taken on the basis of market conditions (if they produce timber), compliance costs, conditions of regulatory environment, norms and attitudes of NGOs leaders, as well as internal and external capacities (Carlsen et al., 2012). But, an additional factor influencing their decisions with respect to firms or public administrations is represented by the attitudes and demands of donors and investors in allocating their funds. In some cases, donors explicitly require an NGO's project to include compliance with certification standards (Brotto et al., 2017).

## 5.2 Forest carbon finance

Adaptation and mitigation of climate change are the foremost and greatest challenges of this century. Forests can play a double role in the fight against climate change. On the one hand, Reducing Emissions from Deforestation and Forest Degradation (REDD) processes account for about 20% of the anthropogenic greenhouse gas (GHG) emissions. On the other hand, new afforestation and reforestations can increase the forest carbon stock, and sustainable forest management can enhance the capacity of forests to sequester and store carbon dioxide from the atmosphere. Thus, forests and forest management play an essential role in reducing global warming by absorbing and storing carbon. Financial mechanisms that put a value on the tons of carbon[4] stored in forests have been adopted at the international level. This mechanism is called carbon offsetting, where an offset is defined as a "credit representing the reduction, avoidance or sequestration of a metric ton of carbon dioxide or GHG equivalent" (Ecosystem Marketplace, 2018). Carbon offsetting represents a cost-effective alternative to emission reduction for governments as well as for companies, organizations and individuals, which can buy and trade these credits to contribute reducing their impact on climate change. This mechanism generated the creation of what is called forest carbon finance. This paragraph provides a snapshot of the markets for forest carbon offsets, while a more in-depth description of carbon offsetting and market functioning is provided in Chapter 8.

The data presented in this chapter is based on the latest Forest Trends' Ecosystem Marketplace reports (Hamrick and Gallant, 2017a; Goldstein and Reuf, 2016; Goldstein et al., 2015). Ecosystem Marketplace has provided market reports since 2008 on the state of forest carbon finance based on data collected from project developers, brokers and retailers, as well as carbon offset accounting registries that track and facilitate the transfer of offsets. According to Hamrick and Gallant (2017a), a total amount of USD 2.8 billion has cumulated to date in forest carbon finance (Table 5.4).

*Table 5.4* Summary of types of forest carbon finance cumulated* (in USD)

| | | |
|---|---|---|
| Market | Voluntary forest carbon offset transactions | 996.6 million |
| | Compliance forest carbon offset transactions | 1,573,9 million |
| Non-Market | Payments for REDD+ programs | 218.0 million |
| Total | | 2,788.5 million |

* *Refers to the total finance known to date since the early 2000s as tracked by Ecosystem Marketplace*

Source: Adapted from Hamrick and Gallant (2017a)

A distinction is made between market and non-market mechanisms. Market mechanisms may be regulatory (compliance markets) or voluntary (voluntary markets). Compliance markets are those created and regulated by mandatory institutional schemes, i.e., Kyoto Protocol mechanisms. Voluntary markets are made up of private businesses, organizations and individual initiatives not motivated by regulatory requirements. In addition, we consider a third cross-cutting category of domestic initiatives that can be either regulated or voluntary-based (see Chapter 8). Non-market mechanisms include payments for REDD+ programs, where the resulting offsets are not tradable and do not appear in a marketplace.

In the biggest compliance carbon market ever created, the Kyoto Protocol, forestry and land use activities play a marginal role. Market mechanisms involving forests under the Protocol are the Clean Development Mechanism (CDM) and the Joint Implementation (JI), whereby industrialized countries can invest in afforestation and reforestation projects in developing countries (CDM) or other industrialized countries (JI) and to get, in exchange, carbon credits[5] to be included in the national carbon accounting systems. As of March 2018, 71 afforestation and reforestation projects have been developed out of a total of more than 8,362 CDM projects, for the equivalent of only the 0.8% of the total credits generated. A similar situation is found for JI, with only 2 approved afforestation and reforestation projects and 1 avoided deforestation project out of 761 for the equivalent of 0.4% of credits. The limited applicability of CDM and JI in the forestry sector is due to the fact that credits only last for few years (credits expire after 5 years and have to be renewed) and methodological complexities in calculating climate benefits, hence favouring only large-scale projects.

Nearly USD 1 billion of forest carbon finance comes from carbon offsets transacted in voluntary markets. The volume of offsets traded in voluntary markets varies greatly from year to year (Table 5.5). However, after peak sales between 2008 and 2010, the market started to contract until 2013. In 2016, the value of offsets transacted (USD 74.2 million) contracted 21% from the previous year (USD 88.4 million) (Hamrick and Gallant, 2017a).

The average price paid for forest carbon offsets in the voluntary market reached 5.2 USD/t$CO_2$e in 2016. However, prices of offsets sold on the voluntary market vary greatly, from less than 0.7 USD/t$CO_2$e to more than 70

*Table 5.5* Voluntary market for forestry and land use offsets in 2014, 2015 and 2016

| *Volume (MtCO2e)* | | | *Value (USD million)* | | | *Average Price (USD/tCO2e)* | | |
|---|---|---|---|---|---|---|---|---|
| 2014 | 2015 | 2016 | 2014 | 2015 | 2016 | 2014 | 2015 | 2016 |
| 23.7 | 18.2 | 14.3 | 128.1 | 88.4 | 74.2 | 5.4 | 4.9 | 5.2 |

Source: Our elaboration based on Hamrick and Gallant (2017a); Goldstein and Reuf (2016); and Goldstein et al. (2015)

USD/t$CO_2$e, depending on a number of factors, including location, type of project, buyer's preferences and presence of additional social and environmental benefits (Hamrick and Gallant, 2017a). Moreover, voluntary carbon offsets can be sold in primary or secondary markets. Primary markets comprise offset sales from project developers or owners to intermediaries or directly to end-buyers; in 2016 these reached a volume of 8.6 Mt$CO_2$e at an average of 5 USD/t$CO_2$e. Secondary markets refer to sales among intermediaries or from intermediaries to end-buyers, which add another 5.7 Mt$CO_2$e at average 5.5 USD/t$CO_2$e. In relation to the locations of forest carbon projects, most of the forest carbon projects are reported to be based in Peru, Brazil, Indonesia and Unites States; while when it comes to voluntary buyers, most of them are from the United States, Netherlands, United Kingdom, France and Germany.

Forestry and land use represents an attractive sector for climate offsetting. One of the key reasons is the potential of forestry projects to deliver additional benefits such as sustainable livelihoods to local communities and biodiversity conservation, which are important elements of corporate social responsibility policies and marketing. Hamrick and Gallant (2017a) reported that in 2016 forestry and land use offset projects represented almost 27% of the transacted volume and 46% of the transacted value in the voluntary market (Table 5.6) (Hamrick and Gallant, 2017b).

Inside forestry and land use activities, the largest activity categories are REDD+, both those that avoid planned causes of deforestation or forest degradation and those that avoid unplanned causes, afforestation and reforestation, Improved Forest Management, agroforestry, urban forestry and grassland management. Such a wide range of activities is possible thanks to the innovation and dynamism permitted in voluntary markets compared to compliance markets. However, the voluntary carbon market is characterized by the absence of public regulations and, as a consequence, credibility and quality standards have gained considerable importance. In fact, 99%

*Table 5.6* Transacted volume, average price and value by project category in 2016

| *Project category* | *Volume (MtCO2e)* | *Price (USD)* | *Value (USD)* |
|---|---|---|---|
| Renewables | 18.3 | 1.4 | 25 million |
| Forestry and land use | 13.1 | 5.1 | 67 million |
| Methane | 5.6 | 1.8 | 10 million |
| Efficiency and fuel switching | 4.5 | 2.9 | 13 million |
| Household devices | 3.4 | 5.2 | 18 million |
| Transportation | 1.9 | 0.3 | 1 million |
| Gases | 1.4 | 5.7 | 8 million |
| Others | 0.5 | 4.0 | 2 million |

Source: Hamrick and Gallant (2017b)

of the credits transacted in the voluntary forest carbon market are certified in accordance with a third-party independent accredited standard. In 2016, the majority of forestry and land use projects were VCS certified (82%), and 73% of these were certified in conjunction with the CCB Standard, a standard focusing on community and biodiversity impacts of forest carbon projects. The next largest share of the market was made by ACR certified offsets (5%) mostly located in the United States. The Gold Standard and Plan Vivo, both with a strong emphasis on co-benefits, accounted respectively for 4% and 2% of the market volume. Other widespread standards are the Climate Action Reserve (CAR) and Social Carbon Standard (Hamrick and Gallant, 2017a) (see Chapter 8).

While historically most of the offsets from forest carbon projects were sold on the voluntary carbon market, the emergence in recent years of domestic market initiatives accepting forest carbon offsets have started to shift this pattern.

Some of these markets functions as compliance markets, and thus are regulated under governmental schemes that require companies to participate. Here, the structure and scope vary greatly among programs. Not all programs allow companies to purchase carbon offsets from forestry projects to meet their obligations, i.e., the European Union Emissions Trading System. Other important regulated domestic carbon programs, such as the California's cap-and-trade and Australia's Emissions Reduction Fund (ERF), only began to accept offsets from forestry and land use projects in 2013. However, despite having a shorter history, over USD 1.5 billion of forest carbon offsets has been transacted in domestic regulatory programs. In 2016, of the USD 551.4 million of compliance forest carbon offsets, USD 509.5 million came only from the Australia's ERF (Hamrick and Gallant, 2017a). In 2016, 13 countries had some form of domestic carbon program accepting trade of forestry and land use carbon offsets. An overview of the main compliance domestic markets evolution in recent years is presented in Table 5.7.

Domestic markets exist also on a voluntary basis, where the government organizes a carbon pricing system but it is up to businesses, organizations and individuals to participate or not, or on a compliance basis, with cap-and-trade systems created and regulated by governments. For example, Japan has the J-Credit system, a consolidated government-managed voluntary carbon offsetting program, as well as the Joint Crediting Mechanism (JCM), a government-administered program that partners with developing countries to invest in carbon offset projects. Also, Mexico has a voluntary domestic carbon trading platform called MexiCO2. The United Kingdom, in addition to participating in the EU ETS, has created the Woodland Carbon Code (WCC), a voluntary domestic offsetting program for forest-based carbon projects (Hamrick and Gallant, 2017a).

*Table 5.7* Overview of main domestic compliance markets in 2014, 2015 and 2016

| | *Contracted Volume (MtCO2e)* | | | *Value (USD million)* | | | *Average Price (USD/tCO2e)* | | |
|---|---|---|---|---|---|---|---|---|---|
| | *2014* | *2015* | *2016* | *2014* | *2015* | *2016* | *2014* | *2015* | *2016* |
| Australia carbon tax/ERF | 4.0 | 60.7 | 68.8 | 70.6 | 588.5 | 509.5 | 17.7 | 9.7 | 7.4 |
| California and Québec cap-and-trade program | 6.1 | 6.5 | 4.1 | 54.7 | 63.2 | Not reported | 8.9 | 9.7 | Not reported |
| New Zealand ETS transactions | 0.4 | 1.3 | Not reported | 2.1 | 10.4 | Not reported | 5.0 | 7.9 | Not reported |
| New Zealand ETS retirements | 0.8 | 10.2 | 15.5 | 2.1 | 45.6 | 188.1 | 3.1 | 4.5 | 10.2 |

Source: Own elaboration based on Hamrick and Gallant (2017a); Goldstein and Reuf (2016); and Goldstein et al. (2015)

## 5.3 Wood energy market

Solid biomass, including fuelwood, charcoal, agricultural and forestry waste, is the most important single source of renewable energy, providing about 6% of the world's total primary energy supply annually (IRENA, 2016). In Africa, wood is used to produce half of all the energy consumed, while one-third of households worldwide (about 2.4 billion people) use wood as their main fuel source for cooking and boiling drinking water (FAO, 2014b). The domestic use of wood for energy is also significant in industrialized countries. According to the FAO (2014b), more than 80 million people in Europe and around 8 million people in North America use fuelwood as their main heating source. The use of woody biomass for electricity is prominent in Europe and North America. Nowadays, around 70% of all solid biomass consumed to produce electricity is used in Europe and the USA (World Energy Council, 2016).

Three types of wood products used for energy dominate the market and the international trade:

- Fuelwood. According to FAO (2014b), the global fuelwood production amounted to 1854 million m3. Fuelwood is mainly consumed at the local level; in fact less than 1% of fuelwood is internationally traded. Official statistics are likely to exclude informal cross-border trade, which represents an important element of the fuelwood market (Lamers et al., 2012a). Informal trade and low-end-value are seen as the major barriers for the certification of fuelwood.
- Woodchips. According to Lamers et al. (2012b), woodchips used for energy represent less than 10% of the global trade of woodchips. The majority of woodchips are destined for pulp, paper and particle board production. Woodchips for energy production are traded mostly to and within the European Union. Woodchips used for residential use are mainly locally produced, whereas international woodchip trade is driven by the industrial sector, where chips are combusted in dedicated energy plants. Woodchips have a multiple end use, hence it is difficult to assess and describe the market size of the certification for woodchips used for energy purposes.
- Wood pellets. The production of pellets has increased since 2000, moving from 2 million tons per year to around 28.9 million tons in 2016 at the global level (AEBIOM, 2017). Pellets are transported over longer distances than fuelwood and woodchips, indeed more than half the global volume of pellets produced in 2016 was traded internationally. In the last decade, the EU has emerged as the world leader in pellet production and nowadays almost half of the pellets produced globally originated in the EU. Other major pellet producing regions are North America, Russia, Brazil and Asia, where the pellet industry is booming

both in terms of production and consumption. The EU is also the major consumer of pellets, with a consumption of 21.7 million tons in 2016 (AEBIOM, 2017).

**Box 5.1 The market for ENplus® certified pellets**

ENplus certification is currently the most diffuse wood energy quality certification scheme at the European level. Initially created in 2010 by the German Pellet Trade Association (DEPV), its implementation in the rest of Europe started from 2011 onwards. In the course of a few years only, ENplus scaled up worldwide, demonstrating the strong demand from the industry and consumers for quality certification in the trade of wood pellets. Currently, 9.2 million tons of pellets are produced under the ENplus certification, representing 67% of the pellets used in the European heating sector (EPC, 2017). The diffusion of the ENplus certification varies from country to country; the countries with the highest pellet certification rate are Germany and Austria, where almost all the pellet production is ENplus certified; also France and Russia have an increasingly important production of ENplus certified pellets. In other countries, such as Sweden and Finland, two of the leading European countries for wood pellet production, the share of ENplus certified pellets remains very low. In fact, the wood energy sector in Scandinavian countries is based on a robust tradition of high quality products, hence, consumers don't feel the necessity for the introduction of new standards. In other countries, such as Hungary, companies operating in the wood energy sector do not have the financial means to invest in the equipment needed to meet the ENplus requirements, and for this reason the number of companies involved in the ENplus certification process remains very restricted. More data and information on the ENplus pellet market can be found in the ENplus Certification Statistical Overview 2017 (EPC, 2017).

## 5.4 Non-Wood Forest Product markets

The market of certified products has boomed in the last two decades thanks to the improvement of customer's awareness toward certified products (ITC, 2009). While market data on agriculture and wood-based materials can be easily found, it is much more difficult to assess the economic value of certified NWFPs. Certification bodies usually provide statistics of the certified area and the number of certified products and value chains, but little information is provided about market values and often these values are related to specific years.

Willer and Lernoud (2017) reports that worldwide the wild collection area (including beekeeping) under organic certification covers 39.7 Mha, which is a very wide area, considering that organic agriculture covers a surface of 50.9 Mha. Most of the wild organic collection land is situated in Europe (about 17.6 Mha), followed by Africa and Latin America (Figure 5.1). The countries with the largest area are Finland (mainly berries), Zambia (beekeeping) and India. Berries, apiculture and medicinal and aromatic plants cover the most important roles, as detailed in Table 5.8.

The market value of organic products was worth USD 81.6 billion in 2015, recording significant annual increments. More than 90% of the global market value for organic certified products is in the USA and Europe. The USA alone holds 53% global market value corresponding to USD 43.3 billion in 2015, while the European market expanded by more than 10% to USD 31.1 billion in 2015. In relation to NWFPs, Censkowsky et al. (2007) estimated the market value at USD 0.8 billion in 2005.

Certified wild forest products market overviews are usually provided either by certification bodies or international associations. For example, FSC reports that certified NWFPs cover a considerable share of world trade today: FSC-certified cork amounts to the 4.6% of the global trade and FSC-certified natural rubber 0.1% (FSC, 2017). Other sources of information are the production and trade data on the conventional products, in which certified commodities represent a minor part. Even using such information,

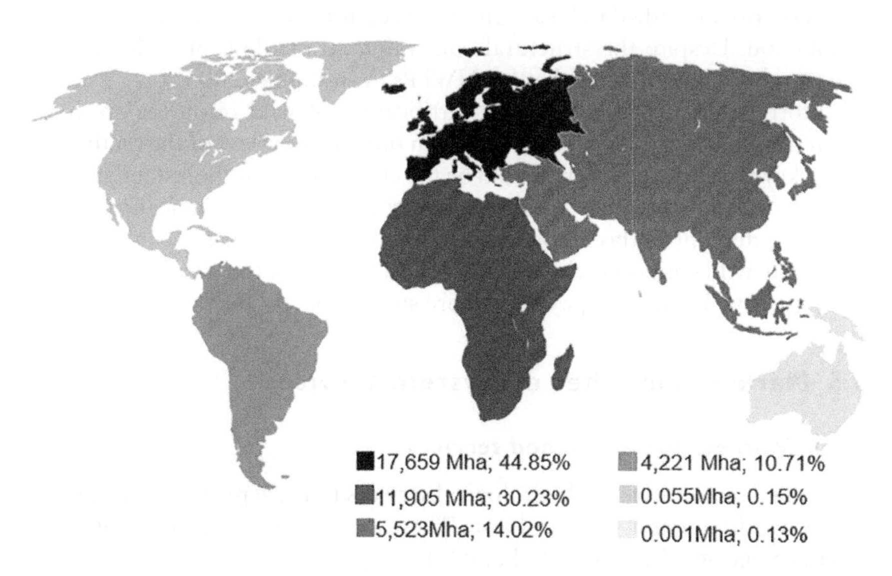


*Figure 5.1* Distribution of wild collection and beekeeping areas under organic schemes, per continent in 2015 (Mha and percentage over the total wild collection area)

Source: Own elaboration based on data from Willer and Lernoud (2017)

*Table 5.8* Wild collection and beekeeping areas by collection type 2015

| *Land use* | *Area (ha)* |
|---|---|
| Apiculture | 6,514,478 |
| Berries, wild | 12,222,218 |
| Fruits, wild | 104,444 |
| Medicinal and aromatic plants, wild | 3,298,249 |
| Mushrooms, wild | 201,006 |
| Nuts, wild | 1,262,415 |
| Oil plants, wild | 964,844 |
| Palm sugar | 1,087 |
| Palmito, wild | 143,867 |
| Rose hips, wild | 170,471 |
| Seaweed | 200,672 |
| Wild collection, no details | 13,183,293 |
| Wild collection, other | 1,096,009 |
| **Total** | **39,363,053** |

Source: Willer and Lernoud (2017)

it is difficult to provide a clear market picture for certified NWFPs. Recently, Pettenella et al. (2014) estimated a global trade value of USD 4.7 billion for the European market of NWFP, showing a structural problem of trade analysis.[6] The three major issues for the market analysis of NWFPs are (i) the complexity of the classification system,[7] (ii) the lack of reporting data in international trade databases, and (iii) the unreliable statistics on NWFPs production. Despite the structural constraints, trade data provides a scale and trend of the global market for NWFPs (Vantomme, 2003).

Unfortunately, standards for the certification of wild resources are fragmented and uncoordinated, resulting in an unclear overview of the retail market. A key issue for the future would be the creation of common standards for the certification schemes of wild forest resources. The communication factor and the externalities (both positive and negative) linked to the consumption of a single wild commodity would raise the attention of the consumers stimulating their willingness to pay for a more sustainable product.

## 5.5 Markets for other ecosystem services

### *5.5.1 Markets for watershed services*

Watershed services are the hydrological services that are provided to a population within a water or river basin. These services are organized into five broad categories (Brauman et al., 2007):

1 improvement of extractive water supply;
2 improvement of in-stream water supply;
3 water damage mitigation;

4 provision of water-related cultural services;
5 water-associated supporting services.

Therefore, Payments for Watershed Services (PWSs) are those payments and/or market schemes that incentivize farmers and forest owners to increase the ecosystem's capacity to provide hydrological services, or to avoid a certain negative impact on specific hydrological services. Payments and market schemes of watershed services are mainly driven by concerns related to water quality issues deriving from agriculture pollution. However, water availability is also a main focus of watershed schemes; some programs are focusing on increasing groundwater recharge, ecosystem retention or ecological river flow, while others deal with water damage caused by flooding and high rainfall variability. Few schemes worldwide reward service providers for water-related cultural services. Two representative examples are the Visitor-Pay-Back schemes within touristic lake and river destinations and the Angling Passport Schemes in the United Kingdom, for restoration of vegetation between the river banks in order to improve access for anglers and fish populations.

In terms of type of ecosystems, Leonardi (2015) shows with a recent study at that EU level that 82% of PWSs mainly target the "built environment" such as agricultural catchments and reforested agricultural lands. Only 50% of the identified PWS target also natural ecosystems such as forests, freshwater bodies and wetlands.[8] Forest-related practices such as Improved Forest Management and reforestation play an important role in providing hydrological services, and they are used by almost half of the EU schemes. In 2015, Ecosystem Marketplace identified approximately 419 active PWS and Water Quality Trading (WQT) programs at the global level (Bennett et al., 2016). In 2008, the baseline year (i.e., the first worldwide study on PWS), about 127 programs were actively receiving payments or transacting water credits. In 2008, the transaction value from all active programs was estimated at €8 billion and increased to €25 billion in 2015. At the EU level, the first report found only 1 active program in 2008, 15 active programs in 2012 and 40 in 2016 (Bennett et al., 2017). It's clear that the general trend is growing, but there has been historical underestimation and lack of data regarding existing and emerging programs in different countries (Bennett et al., 2016). By value, the field is still dominated by national public subsidy programs, which account for more than 90% of funding. Direct investment by water users is still relatively low. More information on water services markets can be found in Bennett et al. (2016) for the global scale and in Bennett et al. (2017) for Europe.

### *5.5.2 Market for biodiversity services*

The market for biodiversity services is led by voluntary and mandatory market-like instruments to manage the biodiversity footprint. The biodiversity compensation mechanisms are known as biodiversity offsets and occur in three different compensation channels: habitat banks, in-lieu fee programs and bespoke compensations. A biodiversity offset is defined as a measurable conservation outcome resulting from actions designed to compensate for significant residual unavoidable biodiversity impacts, arising mostly from infrastructure projects, after appropriate prevention and mitigation measures have been considered. The goal of biodiversity offsets is to achieve local no net loss and preferably a net gain on the ground with respect to species composition, ecosystem function and people's use and cultural values associated with the biodiversity values affected.

Biodiversity services are delivered by different ecosystem types, thus, the bio-market is not limited to forest ecosystems, whether these are certified or not. The habitat or species credits are usually delivered by private landowners or specialized conservation managers (private or public entities), through habitat banks or conservation projects targeting the biodiversity values to offset. The credits generated are evaluated and accredited by environmental agencies to allow their commercialization. In countries like the USA, Canada and Australia, offsets programs allow the legal transfer of offsets liability, creating a suitable environment for entrepreneurs to create and sell environmental services to profit.

Biodiversity credits are the standard units for the trade of biodiversity values. Their calculation is connected to local regulations and to the supervision of agencies in charge of accrediting the accounting and delivery. The application of metrics[9] (i.e., unitary measurements of biodiversity on the ground) results in the credit standard measure, hence their monetary value depends on the type of habitat and species accounted. There have been over 30 types of credits identified among all conservation mechanisms in the world, according to the latest international review published by Ecosystem Marketplace (Bennett and Gallant, 2017).

Driven by an increasing awareness of the accumulated impacts of infrastructure projects and the need for clearer and more efficient offsetting accounting methods, the biodiversity markets are showing a noticeable increase (Madsen et al., 2010). The world's list of Environmental Impact Assessment (EIA)[10] laws is growing as more countries commit to biodiversity conservation. There are nearly 100 biodiversity offsetting programs worldwide are currently active in about 33 countries. The global market size ranges between 2 and over 4 billion USD, resulting in the conservation of about 8.3 million hectares (Bennett and Gallant, 2017).

The price of a single biodiversity credit can vary considerably depending on the biodiversity value to be exchanged (Bennett and Gallant, 2017; Madsen et al., 2010):

- price of wetland credits in the USA ranges from USD 3,000 up to USD 925,000 per credit;

- price of stream credits varies from USD 15 to USD 700;
- price of conservation banking credits ranges from USD 2,500 to USD 140,000.

Given that the broad range of prices, it is difficult to aggregate the market's volume. Nonetheless, the exchange of conservation credits is becoming a more regulated practice as it is being associated with economic and ecological benefits.

The growing market for biodiversity is in need of better and more accurate information. Roughly 80% of the transacted credits are unavailable and not transparent enough to track permitted impacts as well as their compensation. The Ecosystem Marketplace's initiative, SpeciesBanking.com, aims to build the first registry of conservation credits together with the Markit Environmental Registry. This collaboration will pilot a platform for species banking markets in California's Central Valley (Pawliczek and Sullivan, 2011). The platform promises to become a turning point for the global market as the credits would be managed electronically and traded in real time.

Broadly, the voluntary biodiversity market operates similarly to the carbon market. However, there is an important distinction that makes biodiversity credits less interchangeable for international trading: biodiversity impacts are territory-linked. Therefore, in order to achieve suitable compensation (considering the ecosystems and society affected), the bio-credits have to be delivered by land of equal or greater environmental benefit, located in the same impacted ecosystem. In spite of the location anchor, biodiversity credits still share some features with the carbon credits market. Both markets depend on transparency and accountability in order to be reliable and effective for accomplishing the goal of conservation. With this in mind, the Verified Conservation Areas (VCA) initiative unfolded in 2013 from the Green Development Initiative[11] to identify specific areas in need of conservation. Currently, nearly 20 VCAs have been proposed together with their action plan. The goal of this platform is to establish a biodiversity marketplace to scale up the supply of verified conservation outcomes, enabling VCAs to get financial support from international investors. VCAs are in line with the Convention of Biological Diversity[12] and the Convention to Combat Desertification,[13] as they offer finance opportunities and guidance to areas certified under the VCA standard. The auditability of the management plans' implementation, monitoring and evaluation will boost the market of biodiversity services, allowing investors, buyers and donors to benefit from a more transparent scheme to tackle biodiversity loss.

## Notes

1 https://new.usgbc.org/leed
2 https://www.breeam.com/

3 www.changeourfiberboards.com.
4 One ton of carbon dioxide (1 t $CO_2$ e) is equal to one credit of $CO_2$. The abbreviation of equivalent "e" means that the global warming of all greenhouse gases ($CO_2$, $CH_4$, $N_2O$, HFC, PFC and $SF_6$) are translated into $CO_2$ equivalents.
5 "Carbon credits" is in this case used as a synonymous of "carbon offset".
6 Authors focus on NWFPs commonly collected in Europe. The research also includes the importing of such products into Europe. The value is the sum of the values of the HS codes 040900, 060410, 060491, 060499, 070959, 071232, 071233, 071239, 080240, 081040, 081120, 081190, 320110, 320120, 320190 and 450110. The authors applied an empiric coefficient of wild harvested to all the codes based on the information gathered from the project stakeholders (i.e., companies, public agencies, researchers and international institutions).
7 The Harmonized System (HS) codes are the most frequently used and they are ruled by the World Custom Organization (WCO) that revises the coding system according to observations of the states within the World Trade Organization (WTO).
8 A single scheme can reward more than one type of management practice, hence the overall percentages do not necessarily add up to 100%.
9 The credit calculation methods are defined at regional level, nonetheless a general guideline and information source on the metrics and the credits calculation has been published by the Business and Biodiversity Offset Program (BBOP) in the Biodiversity Offsets Design Handbook, available at www.forest-trends.org/documents/files/doc_3101.pdf
10 In North America, USA, Canada and Mexico, there are three laws governing the EIA, respectively. Europe has 13 registered EIA laws, besides the European Directive 2011/92/EU in EIA. For more details about the EIA laws of the world, see www.ecosystemmarketplace.com/pages/dynamic/resources.library.page.php?page_id=8393§ion=our_publications&eod=1
11 The Green Development Initiative international mechanism, funded by the Government of the Netherlands and Switzerland, to explore and review new approaches to finance sustainable landscape management. For more information, see http://gdi.earthmind.net/.
12 See https://www.cbd.int/.
13 See https://www.unccd.int/

## References

AEBIOM (2017). *European Bioenergy Outlook*. AEBIOM Statistical Report 2017. AEBIOM, Brussels.

Auld, G., Gulbrandsen, L.H., McDermott, C.L. (2008). Certification Schemes and the Impact on Forests and Forestry. *Annual Review of Environmental Resources* 33, 187–211.

Bennett, G., Gallant, M. (2017). State of Biodiversity Mitigation 2017 Markets and Compensation for Global Infrastructure Development. Available at www.forest-trends.org/documents/files/doc_5707.pdf#

Bennett, G., Leonardi, A., Ruef, F. (2017). *State of European Markets 2017: Watershed Investments*. Forest Trends Ecosystem Marketplace, Washington, DC.

Bennett, G., Nathaniel, C., Leonardi, A. (2016). *Alliances for Green Infrastructure*. State of Watershed Investment 2016. Forest Trends Ecosystem Marketplace, Washington, DC.

Bouslah, K., M'Zali, B., Turcotte, M.F., Kooli, M. (2010). The Impact of Forest Certification on Firm Financial Performance in Canada and the US. *Journal of Business Ethics* 96, 551–572.

Brauman, K.A., Daily, G.C., Duarte, T.K., Mooney, H.A. (2007). The Nature and Value of Ecosystem Services: An Overview Highlighting Hydrologic Services. *Annual Review of Environment and Resources* 32(1), 67–98.

Brotto, L., Pettenella, D., Pirard, R., Cerutti, O.P. (2017). *Planted Forests in Emerging Economies: Best Practices for Sustainable and Responsible Investments.* CIFOR Occasional Paper, Bogor.

Carlsen, K., Hansen, C.P., Lund, J.F. (2012). Factors Affecting Certification Uptake: Perspectives from the Timber Industry in Ghana. *Forest Policy and Economics* 25, 83–92.

Censkowsky, U., Helberg, U., Nowack, A., Steidle, M. (2007). *Overview of World Production and Marketing of Organic Wild Collected Products* (pp. 1–96). International Trade Centre UNCTAD/WTO, Geneva, Switzerland.

Ebeling, J., Yasué, M. (2009). The Effectiveness of Market-Based Conservation in the Tropics: Forest Certification in Ecuador and Bolivia. *Journal of Environmental Management* 90, 1145–1153.

Ecosystem Marketplace (2018). Glossary. Ecosystem Marketplace, Washington. Available at www.ecosystemmarketplace.com/glossary

EPC (2017). *ENplus® Certification Statistical Overview 2017.* European Pellet Council, Brussels.

FAO (2014). *2013 Global Forest Products Facts and Figures.* Forest Economics, Policy and Products Division. FAO Forestry Department, Rome.

FSC (2016). *Global Market Survey, Forest Stewardship Council Report 2016, Global Development GmbH.* Forest Stewardship Council (International Centre), Bonn, Germany.

FSC (2017). *Market Info Pack 2016–2017.* FSC Global Development GmbH, Bonn, Germany.

Goldstein, A., Neyland, E., Bodnar, E. (2015). *Converging at the Crossroads: State of Forest Carbon Finance 2015.* Ecosystem Marketplace, Washington, DC.

Goldstein, A., Reuf, F. (2016). *View from the Understory: State of Forest Carbon Finance 2016.* Ecosystem Marketplace, Washington, DC.

Hamrick, K., Gallant, M. (2017a). *Fertile Ground: State of the Forest Carbon Finance 2017.* Ecosystem Marketplace, Washington, DC.

Hamrick, K., Gallant, M. (2017b). *Unlocking Potential: State of the Voluntary Carbon Markets 2017.* Ecosystem Marketplace, Washington, DC.

IRENA (2016). Renewable energy benefits: measuring the economics. International Renewable Energy Agency. Abu Dhabi, United Arab Emirates.

ITC (2009). *Consumer Conscience: How Environment and Ethics Are Influencing Exports* (pp. 1–60). International Trade Centre, Geneva, Switzerland.

Kollert, W., Lagan, P. (2007). Do Certified Tropical Logs Fetch a Market Premium? A Comparative Price Analysis from Sabah, Malaysia. *Forest Policy and Economics* 9, 862–868.

Lamers, P., Junginger, M., Hamelinck, C., Faaij, A. (2012a). Developments in International Solid Biofuel Trade, an Analysis of Volumes, Policies, and Market Factors. *Renewable and Sustainable Energy Reviews* 16(2012), 3176–3199.

Lamers, P., Marchal, D., Schouwenberg, P., Cocchi, M., Junginger, M. (2012b). *Global Wood Chip Trade for Energy*. Report Commissioned by IEA Bioenergy Task 40 Sustainable International Bioenergy Trade, Paris.

Leonardi, A. (2015). Characterizing Governance and Benefits of Payments for Watershed Services in Europe. PhD Dissertation at University of Padova. Padova.

Madsen, B., Carroll, N., Kandy, D., Bennett, G. (2011). *Update: State of Biodiversity Markets*. Forest Trends, Washington, DC. Available at www.ecosystemmarketplace.com/reports/2011_update_sbdm

Madsen, B., Carroll, N., Moore Brands, K. (2010). State of Biodiversity Markets Report: Offset and Compensation Programs Worldwide. Available at www.ecosystemmarketplace. com/documents/acrobat/sbdmr.pdf

Owari, T., Sawanobori, Y. (2007). Analysis of the Certified Forest Products Market in Japan. *Holz Roh Werkst* 65, 113–120.

Pawliczek, J., Sullivan, S. (2011). Conservation and Concealment in SpeciesBanking.com, USA: An Analysis of Neoliberal Performance in the Species Offsetting Industry. *Environmental Conservation* 38(4), 435–444.

Pettenella, D., Vidale, E., Da Re, R., Lovric, M. (2014). NWFP in the International Market: Current Situation and Trends. Star-Tree Project Report. Padua, Italy.

Romero, C., Puts, F., Guariguata, M., Sills, E., Cerutti, P., Lescuyer, G. (2013). An Overview of Current Knowledge about the Impacts of Forest Management Certification: A Proposed Framework for Its Evaluation. Occasional Paper 91. Center for International Forestry, Bagor, Indonesia.

Takahashi, T., Van Kooten, G.C., Vertinsky, I. (2003). Why Might Forest Companies Certify? Results from a Canadian Survey. *International Forestry Review* 5(4), 329–337.

Van Kooten, G.C., Nelson, H.W., Vertinsky, I. (2005). Certification of Sustainable Forest Management Practices: A Global Perspective on Why Countries Certify. *Forest Policy and Economics* 7, 857–867.

Vantomme, P. (2003). Compiling Statistics on Non-Wood Forest Products as Policy and Decision-Making Tools at the National Level. *International Forestry Review* 5(2).

Willer, H., Lernoud, J. (Eds.) (2017). The World of Organic Agriculture. Statistics and Emerging Trends 2017. Research Institute of Organic Agriculture (FiBL), Frick, and IFOAM – Organics International, Bonn. Version 1.3 of February 20, 2017.

World Energy Council (2016). *World Energy Resources 2016*. World Energy Council, London.

# Part 2

# Standards and systems of forest products and services certification

Chapter 6

# FSC® Forest Management Certification

*Diego Florian, Ilaria Dalla Vecchia and Mauro Masiero*

This chapter presents the FSC Forest Management (FM) Certification system. The objective is to give an overview of the FM Standards focusing on the most recent version of FSC Principles and Criteria, the standard FSC-STD-01–005 V5–2 (FSC, 2015a). Paragraphs 6.1, 6.2 and 6.3 present the overall framework of the standard, while the main requirements of the standard (e.g., socio-economic impacts, environmental impacts, stakeholder consultation, etc.) are described in paragraph 6.4. The chapter also provides two case studies on the application of the FSC FM Standard: a case study in Canada, which focuses on indigenous people, and a case study in Chile, which focuses on certification of planted forests.

## 6.1 FSC Forest Management Certification: the rationale and key concepts

The FSC system was established in the 1990s and is based on the promotion of environmentally appropriate, socially beneficial and economically viable FM practices around the world. This means that forest management should also ensure "the production of timber, non-timber forest products and ecosystem services and maintenance of the forest biodiversity, productivity and ecological processes" (FSC, 2015a, p. 6), while helping indigenous people and local communities to enjoy forest benefits, not only in terms of financial return, but also in relation to reaching the market with the "the full range of forest products and services for their best values" (FSC, 2015a, p. 6). The FSC system is one of the earliest examples of performance-based standardization processes that brings together non-governmental organizations (NGOs), enterprises and private citizens to promote a forest management model aiming at improved forestry practices that ensure economic benefits for local communities (ISEAL, 2010). FSC certification involves third-party auditing (the audit is carried out by an independent organization, called certification bodies, or CB). Certification bodies are controlled by the accreditation body called Accreditation Services International (ASI). ASI checks each CB's quality system.

In January 2018, the total international FSC-certified area amounted to 198.862 million hectares distributed among 1,533 FM/Chain of Custody (CoC[1]) certificates and followed along the value chain by 33,626 CoC certificates.

## 6.2 Overview of FSC standards

FSC standards for FM Certification are based on the hierarchical structure of the FSC Principles and Criteria (P&C), the international requirements for responsible forestry (FSC, 2015a). There are ten FSC principles that identify responsible, socio-economic and environmental forestry practices. Each principle is composed of specific criteria used to prove whether the principles are fulfilled or not. The first version of the FSC P&C (FSC-STD-01–001) was published in 1994, then amended in 1996, 1999 and 2001. Significant reviews were conducted from 2009 to 2011 that ended in Version 5–1, approved by FSC members at the 7th General Assembly in Seville (Spain) on 12 September 2014. Subject to final revision in 2015 (FSC P&C V5–2), FSC P&C are valid all around the world, for all forest types and management areas, including natural forests, planted forests and other vegetation types. Certification refers to products and services that cover wood and Non-Timber Forest Products as well as ecosystem services such carbon sequestration, water flooding regulation and biodiversity conservation and protection measures. From a legal perspective, FSC is supporting responsible forest initiatives already in place around the world, ensuring that FM is in line with, and sometimes exceeding, local, national and even international regulations.

## 6.3 International Principles and Criteria and national standards

The FSC P&C cannot be used directly in the forest for the evaluation and certification of responsible forest stewardship; an additional approved set of indicators adapted to national, regional or local conditions is needed. Before the new FSC FM standard version (5–2), National Initiative, Regional Offices or National Standard Working Groups were responsible for the definition of National Standards and Indicators for FSC, according to the standard FSC-STD-60–006 V1–2 (FSC, 2009a). As of today, a set of FSC International Generic Indicators (IGIs) provides a homogeneous and structured transfer process to facilitate forest managers, stakeholders and certification bodies to apply the internationally recognized requirements (principles and criteria) at the regional and local levels. The IGIs are collected in the standard FSC-STD-60–004 V1–0 EN and are effective from September 2015.

Additional information concerning the transfer of IGIs for the creation of National Forest Stewardship Standards (NFSS) is provided in the procedure

FSC-PRO-60–006 V1–0 (FSC, 2015b). The aim is to ensure the consistent application of the internationally recognized requirements across the globe, in a coherent and pragmatic way. Groups of technical experts, defined as National Standard Development Groups, are asked to develop an NFSS taking the requirements of IGIs, a minimum number of stakeholder consultations and field pilot tests into consideration. IGIs will replace certification bodies' *ad interim* standards, where national standards do not exist; while for countries with an approved NFSS, or at minimum a draft version, IGIs will be transferred and harmonized at the national level (FSC, 2015b); in other words, IGIs are defining the FSC forest management requirements on the ground. FSC P&C adapted at the national level aim at recognizing legal, social, ecological, cultural and geographical territorial level diversities. The scale, intensity and risk of management actions may vary from area to area, generating different impacts; the need for a more flexible system of standard development has been satisfied through the creation of the Scale, Intensity, Risk Guideline (SIR), to be fulfilled during the process of IGI adaptation, adoption, drop or addition of indicators (FSC, 2016). In relation to the transfer process, adaptation of the last version of FSC P&C was in December 2015, when NFSS became effective.

## 6.4 FSC Forest Management Certification: main requirements

FSC P&C cover all forest management activities at different Forest Management Unit (FMU) scales; requirements refer to the socio-economic and environmental impacts, specifying the legality of forest management activities and the responsibility for ensuring compliance within the FSC P&C. The people/entities acting for the direct and indirect management activities within the FMUs must comply with the locally valid FSC-approved national standard. In the case of one or more non-compliances, the organization must apply specific corrective actions as prescribed by auditors during the main assessment or surveillance visits. In other words, FSC P&C and indicators are based on performance-based standards, which rely on the level of forest management satisfaction and the consequences of failing those requirements (FSC, 2009e).

### *6.4.1 Environmental impacts*

Environmental impacts are addressed by Principles 6 and 10. Impacts are any potential adverse effects within the implementation of FM activities. In more detail, the organization is required to "conserve and/or restore ecosystem services and environmental values of the Management Unit, and shall avoid, repair or mitigate negative environmental impacts" (FSC, 2015a, p. 14), and the assessment shall be conducted in relation to the scale, intensity and risk of the management activities. An example of this is the protection and/or restoration of native ecosystems. The Conservation Area Network should at least reach 10% of the FMU area, a value that can be increased in the case of large-scale FMUs, intensively managed forests and landscapes with scarce protection areas.

Potential negative impacts must be considered before starting the utilization of the FMUs, providing a set of prevention and mitigation activities. Specific attention is given to the identification and protection of (i) rare and threatened species and their habitats (protection, conservation and connectivity areas) (Criteria 6.4); (ii) representative sample areas of native ecosystems or areas within the FMUs to be restored to more natural conditions (Criteria 6.5); (iii) naturally occurring native species/genotypes for the prevention of biological diversity loss (Criteria 6.6); and (iv) water courses, water bodies, riparian zones and their connectivity (Criteria 6.7). All FMUs should be planned incorporating a landscape approach to maintain and restore "a varying mosaic of species, sizes, ages, spatial scales and regeneration cycles appropriate for the landscape values in that region, and for enhancing environmental and economic resilience" (FSC, 2015a, p. 15). Restrictions are set in case of forest conversion; natural forests converted to plantations or other non-forest land uses (see Box 6.1) cannot be certified if conversion took place after November 1994. Moreover, activities such as management planning, harvesting and regeneration techniques, silviculture practices, absence of genetically modified organisms (GMOs), use of fertilizers and disposal of waste shall be conducted according to "the Organization's economic, environmental and social policies and objectives and in compliance with the Principles and Criteria collectively" (FSC, 2015a, p. 19). This means that the organization works in line with FSC requirements, assessing management risks and implementing activities that reduce potential negative impacts caused by human and natural factors, proportionate to the scale, intensity and risks of the management activities.

### Box 6.1 The Chilean case study

The total international area of planted forests is increasing, both for protective and productive purposes, and generating a whole set of impacts that are increasingly considered by society and investors. Chile is an interesting case as, out of the total international area of certified forests, it is the country with the highest number of certified planted forests. Since 2010, Chile is the only country with specific national standards for plantations. A recent study (Masiero et al., 2015) has shown that the FSC National Standard for planted forests in Chile is more stringent and credible than other schemes (PEFC and LEI).[2] The study compared many standards with a theoretical ideal standard for planted forests. Particular attention was given to the fact that the indicators of the FSC Chile Standard incorporate around 50% of the theoretical ideal standard's indicators, and it is defined as the basic tool for conducting a gap analysis. The FSC Chile Standard is the best-performing in terms of policy and planning for sustainable and multifunctional forest management; the FSC Chile Standard is the best-performing in terms of policy and planning for sustainable and multifunctional forest management and for biodiversity and ecological processes protection and maintenance. There were only a few weaknesses found in the aforementioned study of the FSC Chile Standard: the protective functions of forests (e.g., forests that shall be protected, maintained and where possible strengthened), soil protection, water management and carbon cycle maintenance.

### *6.4.2 Social impacts*

Social impacts rely on P&C from the first to the fourth principles. The first principle applies to all socio-economic and environmental aspects, as the organization shall "comply with all applicable law, regulations and nationally ratified international treaties, convention and agreements" (FSC, 2015a, p. 10). Social impacts within a specific FMU are strictly related to the "social and economic well-being of workers" (FSC, 2015a, p. 11), according to the ILO Declaration on Fundamental Principles and Rights at Work (1998), based on the eight ILO Conventions. The promotion of gender equality in the workplace, the implementation of safety practices and training activities and the fair compensation for any occupational employment risks are some of the basic requirements for FM Certification, oriented to increase engagement processes and training opportunities within and outside the FMUs. Principle 3 dealing with indigenous peoples' rights (see Box 6.2) is focused on the identification of indigenous peoples' "legal and customary rights of ownership, use and management of land, territories and resources affected by management activities" (FSC, 2015a, p. 12). As stated by Article 12 of the United Nations Declaration on the Rights of Indigenous People (UN, 2008), "indigenous people have the rights to maintain, protect and have access to their religious and cultural sites" and need to be protected and managed in a proper way within or outside the FMUs. Indigenous people are consulted through FPIC before any utilization activities take place, in order to ensure that they "can understand and be understood in political, legal and administrative proceedings" (Art. 13, part 2). Indigenous peoples' engagement is also essential to mitigate any kind of adverse impacts deriving from FM Operations, and the organization shall provide appropriate compensation mechanisms if these impacts occur. According to Principle 4, organizations shall also identify local communities within, or affected by, the FMUs (FSC, 2015a). Their legal and customary rights must be recognized, and where needed FPIC is used by delegations of local communities to check the FM activities. Opportunities of employment, training and other services must be provided to locals, to enhance and/or maintain their economic well-being, always proportionate to the scale, intensity and socio-economic impacts and risks connected to management activities. Any forms of conflicts or disputes with local communities must be solved through *ad hoc* mechanisms, providing fair compensation in relation to the impacts caused.

**Box 6.2 Indigenous people in Canadian forests**

The issue of indigenous peoples' rights and their contribution towards the development of rational P&C is strongly part of the FSC system. In particular, FSC Principle 3 recognizes the "culturally appropriate engagement with Indigenous people" in order to identify land use rights, including the access to forests and related services, their customary rights and legal obligations within the Forest Management Unit. Free, prior and informed consent (FPIC) must be granted to the organization before any management activities take place. Under this framework, FSC Canada became partners with the National Aboriginal Forestry Association (NAFA)[3] to promote and support the involvement of Aborigines in forest management and related market opportunities. In December 2014, forest managers, technicians, academics, lawyers and industry representatives together with the chiefs of the First Nations discussed the environmental, cultural and economic values of their lands and how to include these values into responsible forestry. The president of Canada states

> First Nations are key players in the global forest economy with millennia of experience in the successful management of diverse forest homelands. We see this event as a key component in our engagement with First Nations to rigorously apply free, prior and informed consent into FSC's Forest Management Standard.
>
> (FSC, 2015c)

### *6.4.3 Economic impacts*

Forest economic benefits are provided in line with Principle 5, "enhancing long-term economic viability and the range of environmental and social benefits" (FSC, 2015a, p. 14); in other words, all harvesting activities related to specific products and services must be conducted with sustainable levels, in relation to the scale and intensity of management activities. According to Criteria 6.2, the organization shall harvest products and services from the FMU at a level that can be permanently sustained and ensure long-term provision of the products. It this way, FSC markets and promotes what nature provides rather than just what markets demand, generating added value from wood and Non-Wood Forest Products. Further standard improvements are going in the direction of multifunctional forest services. Different ecosystem services, such as biodiversity protection, water cycle regulation, soil conservation and carbon storage, are likely to be monitored and evaluated in the near future. The ForCES project,[4] for example, improves and promotes sustainable forest management for a range of ecosystem services (FSC, 2018). Another relevant issue is the proximity of production and processing as a strategy to increase long-term sustainability and viability of forest products: "local processing, local services, and local value adding to meet the requirements of the Organization where these are available, proportionate to scale, intensity and risk" (FSC, 2015a, p. 14).

### *6.4.4 Management systems: planning and monitoring*

Planning and monitoring information are provided within Principles 7 and 8. The main core issues related to the management system are the development, implementation and follow up of a forest management plan (FMP), together with the implementation of the organization's management activities, policies and objectives. The FMP describes the natural resources as well as the way that FSC economic, social and environmental requirements are met. The FMP includes verifiable targets and integrates updates and revisions from the results of monitoring and evaluation activities. A summary of the management plan should be publicly available free of charge. Management activities "should proactively and transparently engage affected stakeholders in its management planning and monitoring processes". Monitoring processes are based on changing environmental, social and economic conditions, from forest management along all value chain steps, to the trademark uses. A tracking and tracing system must be proportionate to the scale, intensity and risk of management activities, in order to demonstrate "the source and volume in proportion to projected output for each year, of all products from the Management Unit that are marketed as FSC certified" (FSC, 2015a, p. 17).

### *6.4.5 Stakeholder consultation*

Stakeholders include "any person, group of persons, or entity that has shown an interest, or is known to have an interest, in the activities of a Management Unit" (FSC, 2015a, p. 20). Stakeholders provide a significant contribution to the FSC system, based on a democratic, transparent and inclusive structure and governance. If interested or affected by a range of management activities, stakeholders are invited to give feedback in relation to both national and international standard-setting processes. In both cases, they are asked to provide comments in relation to the first draft provided either by working groups or technical advisors. Before the approval by the FSC Policy and Standard Unit, at least two consultation sessions are held. Stakeholders' inputs are to be considered also in relation to certification processes: during the main evaluations and annual verification by certification bodies, contributions are given by individuals or groups of people affected or interested in the forest to be certified. Also, for the organization there is an initial consultation phase followed by a continuum process of consultations with stakeholders or third/interested parties. The standard FSC-STD-20–006 V3–0 provides the requirement for the range of stakeholders to be informed, the information provided and the methods to analyse and evaluate stakeholders (FSC, 2009b). A list of potential key informants includes the FSC national partner, the local and national forest service, national or international NGOs actively involved in forestry issues, representatives of indigenous peoples and/or local communities, labour organizations and contractors within the forestry operations assessed. Certification bodies shall also inform stakeholders at least 6 weeks before the main evaluation processes start, either by direct e-mail, individual or collective interviews and public announcements using any available means. Consultations shall be summarized and publicly available, with a systematic presentation of stakeholder comments and a description of follow-up actions.

### 6.4.6 Management of High Conservation Values

The definition of High Conservation Values (HCVs) has been modified according to the last version of P&C (FSC, 2015a). It has become *a biological, ecological, social or cultural value of outstanding significance or critical importance* (Brown et al., 2013). Before, it was just referring to environmentally significant conservation areas within FMUs. There are six categories of HCVs (Table 6.1).

Therefore, HCVs include not only environmental-related characteristics to be protected but also community needs and cultural values essential for local or indigenous populations within the FMUs of interest. According to FSC requirements, HCVs must be identified within the forest management plan, assessed by stakeholder consultation and properly monitored in terms of implementation and effectiveness of measures carried out by the organization. In order to manage HCVs, the organization is required to identify potential threats and understand which approaches are needed to maintain, protect or enhance HCVs. Monitoring activities translate management objectives into practices, through the use of the precautionary approach, proportionate to the scale, intensity and risks of the management activities, and the level of engagement of interested or affected stakeholders.

*Table 6.1* High Conservation Values classification

| *High Conservation Values* | *Contents* |
|---|---|
| HCV 1 | Species diversity. Concentrations of biological diversity including endemic species and rare, threatened or endangered species, that are significant at global, regional or national levels. |
| HCV 2 | Landscape-level ecosystems and mosaics. Intact forest landscapes and large landscape-level ecosystems and ecosystem mosaics that are significant at global, regional or national levels, and that contain viable populations of the majority of the naturally occurring species in natural patterns of distribution and abundance. |
| HCV 3 | Ecosystems and habitats. Rare, threatened or endangered ecosystems, habitats or refugia. |
| HCV 4 | Critical ecosystem services. Basic ecosystem services in critical situations, including protection of water catchments and control of erosion of vulnerable soils and slopes. |
| HCV 5 | Community needs. Sites and resources fundamental for satisfying the basic necessities of local communities or indigenous peoples (for livelihoods, health, nutrition, water, etc.), identified through engagement with these communities or indigenous peoples. |
| HCV 6 | Cultural values. Sites, resources, habitats and landscapes of global or national cultural, archaeological or historical significance, and/or of critical cultural, ecological, economic or religious/sacred importance for the traditional cultures of local communities or indigenous peoples, identified through engagement with these local communities or indigenous peoples. |

Source: FSC (2015a)

## 6.5 Special tools and approaches for smallholders

Smallholders are defined as forest owners or producers that either manage small areas of forest or harvest timber at a low intensity. Smallholders are either part of a community of forest managers or agree as a group to let a third party manage the forest on their behalf. They are, for example, small woodlot private owners, harvesters of Non-Timber Forest Products and community forest operators (FSC, 2009c). FSC provides specific certification tools for smallholders – group certification (6.5.1) and Small and Low Intensity Managed Forests (SLIMF) procedures (6.5.2) – reducing costs and overcharges, while providing access to a series of direct and indirect benefits. In addition, FSC promotes the products derived from smallholders to favour the access of new markets, the possibility to sell products with a premium price and the financial support received from public and private institutions.

### *6.5.1 Group certification*

A group certification includes a group of forest owners or producers that have come together to obtain certification (FSC, 2009d). Within the group, responsibilities differ, based on "a group entity", which may be an individual, a cooperative body, an owners' association or a similar entity. The group entity represents forest members and is in charge of ensuring the compliance of standards to certification bodies. In other words, the group entity holds a single certificate, on behalf of the group members, and it is responsible for management activities. This approach is implemented to reduce certification costs and management operations overload, as the group entity is the one providing information, developing training activities and providing technical support to all members. In this way, costs are reduced, as well as the activities for certification preparation. In terms of group size, there are no restrictions related to the number of members or in relation to hectares covered. If there are homogeneous conditions at the national level, certification is referred to as multi-sites group (within the same country). On the other hand, the monitoring activities within the control system need to be defined. The group entity is asked to visit a sample group of members regularly (at least annually) to ensure compliance with the applicable standards. The benefits of having a group certification structure are connected to the potential opportunities reached by the collective scale of production to enter the markets and to obtain better prices.

### *6.5.2 Small and Low Intensity Managed Forests*

SLIMFs are defined as (i) small in area, i.e., less than 100 hectares (or up to 1,000 ha in some cases);[5] (ii) having a low intensity harvesting rate, i.e., less than 20% of the mean annual increment (MAI) and either an annual harvest from the total production forest area that is less than 5,000 m3 or an average annual harvest from the total production forest that is less than 5,000 m3/yr during the period of validity of the certificate; and (iii) forests that are managed exclusively for Non-Timber Forest Products (NTFPs) excluding plantations of NTFPs (such as oil palm or cocoa plantations) (FSC, 2004). FSC recognizes that costs related to third-party independent certification are relatively high for small enterprises. Moreover, potential environmental, social and economic impacts of small and low intensity forests and related enterprises are relatively low, looking at their potential adverse effects on the ground. The idea of implementing streamlined certification procedures is connected to the need to reduce the costs of the certified organization or entity (FSC, 2015c). Moreover, SLIMFs' definition should be overtaken by certification bodies and always where possible, by existing national offices.

## 6.6 Current developments

The need to adapt the new version of P&C V5, approved in 2012, in a global and consistent way has moved attention to the set of International Generic Indicators (IGIs). National initiatives or national offices, where they exist, or otherwise certification bodies, are responsible for transferring IGIs to national realities. National Standard Development Groups are being created or re-activated in order to add, adopt, adapt or drop generic requirements (IGIs) into a specific geographical scale. This process, connected with a bottom-up and top-down flow of information, is facilitated by the Transfer Template and Matrix (FSC, 2013), helping developers of standards meet the scale, intensity and risks within the geographical scope defined (at the local, national and international levels). The IGI transfer will harmonize the international situation, which currently lacks real coordination between national standards, and will have relevant impacts on the way it will harmonize different types of standards, defined per region or forest type (e.g., plantations, natural forests, SLIMF, etc.). The use of IGIs will create a common, internationally recognized baseline on which national indicators are developed, whether NFSS are in place or not. This will lead to an improved quality of national standards, supporting a faster and more efficient approval process and bringing global consistency to the FSC System.

## Notes

1 FSC Chain of Custody certification allows companies to label their FSC products, which in turn enables consumers to identify and choose products that support responsible forest management. For more information visit ic.fsc.org.
2 PEFC is the Pan-European Forest Council, www.pefc.it; LEI is the Lembaga Ekolabel Indonesia, www.lei.or.id.
3 The overall mission of NAFA is to promote Aboriginal involvement in forest management. NAFA is committed to "multiple-use" forestry in order to satisfy the multitudes of community needs, such as the protection of wildlife and traditional foodstuff habitat, clean and adequate supply of water, establishment of forest areas for recreational activities and tourism, spiritual and cultural uses and the production of timber and Non-Wood Forest Products (more info at www.nafaforestry.org).
4 The Forest Certification for Ecosystem Services (ForCES) project evaluates the changes needed by the FSC system to accommodate a market-based approach for ES. The project is taking place in ten different forest sites for a four-year period. For more information, please visit the website www.forces.fsc.org.
5 Cases are decided by the FSC-accredited national initiative for the country concerned, or in countries in which there is no FSC-accredited national initiative when this has demonstrated the broad support of national stakeholders in the country concerned.

## References

Brown, E., Dudley, N., Lindhe, A., Muhtaman, D.R., Stewart, C., Synnott, T. (eds.) (2013, October). Common Guidance for the Identification of High Conservation Values. HCV Resource Network, Oxford.

FSC (2004). SLIMF Eligibility Criteria FSC-STD-01–003 (V1–0). Bonn, Germany.

FSC (2009a). Process Requirements for the Development and Maintenance of National Forest Stewardship Standards. FSC-STD-06–006 V1–2 EN. Forest Stewardship Council (International Centre). Bonn, Germany.

FSC (2009b). Stakeholder Consultation for Forest Evaluations.FSC-STD-20–006 V3–0 EN. Forest Stewardship Council (International Centre). Bonn, Germany.

FSC (2009c). *FSC User-Friendly Guide to FSC Certification for Smallholders*. FSC Technical Series 2009. FSC International Center GmbH, Oaxaca, Mexico.

FSC (2009d). FSC Standard for Group Entities in Forest Management Groups. FSC-30–005 V1–0 EN. Bonn, Germany.

FSC (2009e). Forest Management Evaluation. FSC-STD-20–007 (V 3–0).

FSC (2013). Transfer Matrix for the Transfer of Forest Stewardship Standards to the Principles & Criteria Version 5. FSC-STD-01–001 V5. Bonn, Germany.

FSC (2015a). FSC Principles and Criteria for Forest Stewardship. FSC-STD-01–001 V5–1 EN. Forest Stewardship Council (International Centre). Bonn, Germany.

FSC (2015b). Development and Transfer of National Forest Stewardship Standards to the FSC Principles and Criteria Version 5. FSC-PRO-06–006 V1–0 EN Forest Stewardship Council (International Centre). Bonn, Germany.

FSC (2015c). Newsroom: News and Topics around Our Activities. https://ca.fsc.org/newsroom.239.365.htm, last consultation in January 2015.

FSC (2015d). Strategy of FSC International. FSC International. Bonn, Germany.

FSC (2016). Scale, Intensity and Risk Guidelines for Standard Developers. FSC-GUI-06–002 V1–0 EN. Forest Stewardship Council (International Centre). Bonn, Germany.

FSC (2018). ForCES: Final Report. FSC International. Bonn, Germany.

ILO (1998). *ILO Declaration on Fundamental Principles and Rights at Work*. International Labour Organization. Geneva, Switzerland.

ISEAL (2010). *Setting Social and Environmental Standards v.5–0 EN*. ISEAL Code of Good practice. ISEAL Alliance, London, UK.

Masiero, M., Secco, L., Pettenella, D., Brotto, L. (2015). Standards and Guidelines for Forest Plantation Management: A Global Comparative Study. *Forest Policy and Economics*. http://dx.doi.org/10.1016/j.forpol.2014.12.008

UN (2008). United Nations Declaration on the Rights of the Indigenous People (2007). www.un.org/esa/socdev/unpfii/documents/DRIPS_en.pdf

Chapter 7

# FSC® Chain of Custody certification

*Mauro Masiero and Diego Florian*

This chapter provides an overview of Chain of Custody (CoC) certification according to FSC standards. After introducing the rationale and key concepts of CoC, the main standards are presented and described. Special attention is paid to the document FSC-STD-40–004 V3–0, with regards to both system (or universal) requirements and systems for controlling FSC claims, but additional standards for Controlled Wood, reclaimed materials and multisite/group certification are briefly presented as well. Finally, FSC labelling requirements are introduced. Chapter 7 is complemented by Chapter 14, which deals with the FSC standards for CoC auditing.

## 7.1 Chain of Custody certification: the rationale and key concepts

FSC standards define Chain of Custody (CoC) as

> the path taken by products from the forest, or in the case of recycled materials from the moment when the material is reclaimed, to the point where the product is sold with an FSC claim and/or is finished and FSC-labelled. The CoC includes each stage of sourcing, processing, trading and distribution where progress to the next stage of the supply chain involves a change of product ownership.
>
> (FSC, 2016, p. 26)

An organization shall hold a valid CoC certificate if they want their products to be FSC claimed, i.e., if they want to sell a product/material as certified or to label and promote it by using FSC trademarks (see 7.3.1 and 7.6 below for details). Implementing a CoC system allows an organization to control their processing system and show their customers the origin of forest-based materials (wood and non-wood) in their products.

The concept of CoC implies an unbroken chain of certified organizations covering every change in legal ownership of the product from the certified forest up to the point where the product is finished or sold to retail (FSC, 2016). A CoC system can be implemented in the form of a joint forest management (FM) and CoC certificate (FM/CoC) or of a sole CoC certificate (Figure 7.1).

In the first case, CoC requirements for forest managers are defined by Criterion 8.5 of FSC-STD-01–001 version 5–2, according to which forest managers are requested to develop, implement and maintain a tracking and tracing system of forest products within the Forest Management Unit (FMU) until it reaches the forest gate.[1] The system shall include complete, correct and up-to-date records of quantities of FSC-certified products (i.e., harvested, stored and sold quantities) and effective measures to allow the identification and separation of FSC-certified material from non-certified material at all stages prior to transfer of ownership. Identification and separation measures shall be defined within specific procedures and might include, for example, dedicated storage areas, hammer brands, paint spots, tags, etc. If products from certified FMUs are mixed with non-certified material and/or FSC-certified material from other FSC Certificate holders, a separate CoC certification is needed as well. This is also the case for primary (e.g., sawmill) or secondary (e.g., wood panel factory) processing facilities associated with the forest management enterprise – they need to be managed according to relevant CoC standards presented in the following paragraphs. A separate CoC certificate is not required for log cutting or de-barking units and small portable sawmills associated with the forest management enterprise.

When a single CoC certification is needed, i.e., in the above-mentioned cases and, more in general, for all operations occurring downstream of the certified FMU (Figure 7.1), reference shall be made to specific CoC standards presented in paragraph 7.2 and the following paragraphs.

## 7.2 Overview of FSC standards for Chain of Custody certification

Requirements for FSC CoC certification are laid down by a large set of standards and other documents summarized in Figure 7.2. CoC standards belong to the CoC certification program. They can be distinguished into

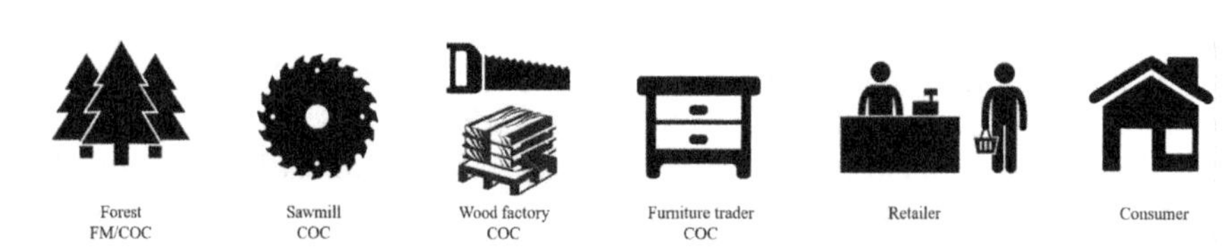


*Figure 7.1* Forest supply chain and FSC Chain of Custody certification

Source: Own elaboration

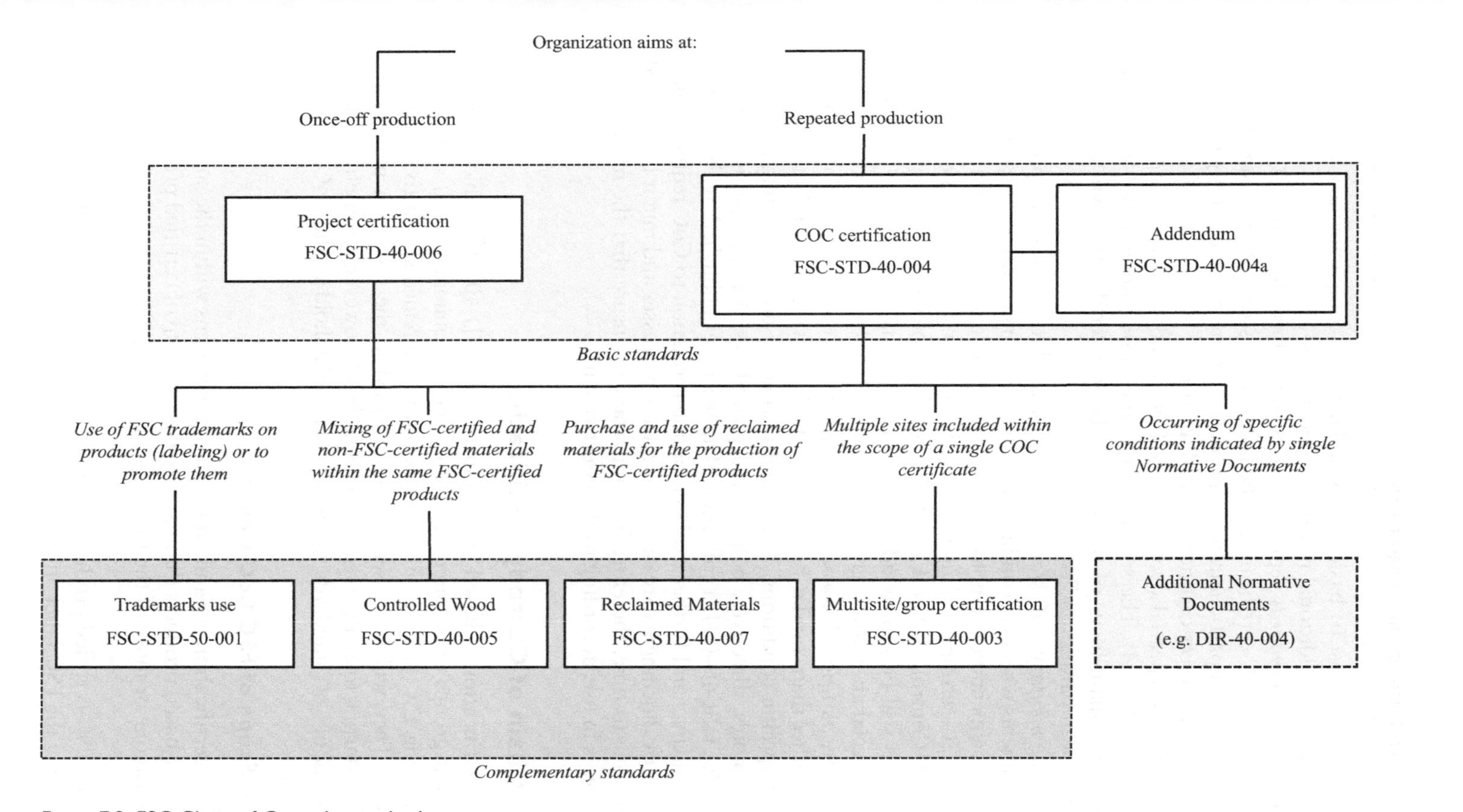


*Figure* 7.2 FSC Chain of Custody standards

Source: Own elaboration

three main groups: (i) basic CoC standards; (ii) complementary CoC standards; and (iii) additional normative documents.

*Basic CoC standards* represent the essential basis for any CoC certification, i.e., no CoC certificate can be issued unless one of these standards is implemented. In particular, if the product is being produced only once, the standard for project certification shall be used (FSC-STD-40–006), while if production is repeated over time, the CoC standard shall be implemented (FSC-STD-40–004). The second case is the most common and will therefore be the main focus of this chapter. Project certification will be shortly addressed in paragraph 7.7.

*Complementary CoC standards* shall be used to integrate the previous set of documents whenever conditions for their implementation occur; otherwise, their implementation is not required. For example, if an organization wants to mix FSC-certified and non-FSC-certified virgin materials to produce one product, Controlled Wood standards shall be implemented. Complementary standards will be shortly addressed in paragraphs from 7.4 to 7.7. Among them, special attention is given to trademark use (paragraph 7.8).

Finally, *additional normative documents* include FSC normative documents other than standards that integrate the previous sets and shall be used under specific conditions. Additional normative documents might include advice notes, directives, policies and procedures. An example is given by Directive DIR-40–004, a collection of advice notes published by FSC to provide guidance and further explanation/interpretation to CoC requirements.

All CoC documents are subject to review processes and might be updated from time to time, therefore it is important to ensure that the most recent version is in use, according to the effective date.

## 7.3 Chain of Custody certification

CoC certification is based on document FSC-STD-40–004, which includes different groups of requirements: (i) universal requirements; (ii) systems for controlling FSC claims; (iii) supplementary requirements; and (iv) eligibility criteria. They will be presented and analysed in the following paragraphs. Before starting to analyse requirements in detail, however, it is useful to clarify the scope of CoC standards, i.e., who is required to hold a valid CoC certificate.

### *7.3.1 Scope of FSC CoC standards*

A CoC certification is required for all organizations within the supply chain of forest-based products that have legal ownership of certified products and perform one or more of the following activities (FSC, 2016):

a sell FSC-certified products with FSC claims on sales documents;[2]
b apply the FSC label on-product;

c process or transform FSC-certified products (e.g., manufacturing, repackaging, relabelling, adding other forest-based components to the product);
d promote FSC-certified products, except finished and FSC-labelled products that may be promoted by non-Certificate holders (e.g., retailers).

When the above-mentioned conditions exist, certification is required independently from physical possession (and processing) of the products. For example, a wood or paper broker that has legal ownership of certified products and wishes to sell them as certified must hold a valid CoC certificate.

Organizations that do not perform the activities described above are exempt from CoC certification, including, for example, retailers selling to end users, end users of FSC-certified products and organizations providing services to certified organizations without taking legal ownership of the certified products (e.g., contractors).

### *7.3.2 System requirements*

CoC system requirements include requirements for (i) management systems, (ii) material sourcing, (iii) material handling, (iv) material and product records, (v) sales and (vi) compliance with timber legality legislation. They will be briefly presented in the next sub-paragraphs after a brief definition of CoC scope.

#### *7.3.2.1 CoC scope*

The scope of a CoC certificate is defined by (i) sites (see 7.6), (ii) product groups, (iii) activities and (iv) certification standards against which an organization is audited (Figure 7.2) (FSC, 2016). A product group is a "group of products specified by the organization, which share basic input and output characteristics and thus can be combined for the purpose of control of FSC output claims and labelling" (FSC, 2016, p. 29). Each product group shall belong to the same product type according to the classification defined in FSC-STD-40–004a and be controlled according to the same FSC control system (see 7.3.3). For product groups established under the percentage and/or credit system (see 7.3.3), two additional conditions apply: (i) all products shall share the same conversion factor indicating the efficiency of the product transformation process under the form of output/input ratio[3] (conversion factors are important for volume balance and in/out material accounts); and (ii) all products shall be made of the same input material (e.g., oak lumber) or same combination of input materials (e.g., veneered fibreboards made of a combination of fibreboards and veneers of equivalent species in terms of price, grade and appearance).

An FSC-certified organization shall define and update a product group list including all the products the organization is willing/able to produce and trade as FSC-certified. For each product group the following elements must be indicated (FSC, 2016) and will be made publicly available via the FSC database: (i) product type (as of FSC-STD-40–004a), (ii) the applicable FSC claims for the outputs (Table 7.1) and (iii) wood specie(s) when information on species composition is commonly used to designate the product characteristics.[4]

### *7.3.2.2 Management systems*

An FSC-certified organization is required to develop and implement a management system for managing the CoC. In particular, this includes (i) the appointment of a person/figure (the CoC manager) who has full responsibility for compliance with all applicable FSC standards, (ii) the implementation and maintenance of procedures covering all applicable FSC requirements (e.g., sourcing, sales, labelling, training, records, etc.), (iii) the definition of roles and responsibilities among staff, (iv) training needs and programs according to roles and responsibilities and (v) 5-year records for all relevant documentation and evidence. All FSC-certified organizations shall commit to comply with the values of FSC defined in the Policy for the Association of Organizations with FSC (FSC, 2011) and show their commitment to occupational health and safety. As for the latter, this implies a representative for occupational health and safety is appointed, procedure(s) for occupational health and safety are developed and implemented and relevant staff is trained on health and safety procedures (FSC, 2013). Additional requirements include the development and implementation of specific procedures for (i) managing complaints received on the organization's conformity to the CoC requirements and (ii) ensuring that any non-conforming products are identified and controlled to prevent their unintended sale and delivery with FSC claims. The latter shall also include specific procedures to ensure non-conforming products are detected after they have been delivered.

### *7.3.2.3 Material sourcing*

The organization shall keep records of all suppliers providing materials to be used as inputs for FSC product groups. Records shall include at least (i) supplier's name, (ii) the supplier's FSC CoC or FSC Controlled Wood (see 7.4) codes (if any), and (iii) materials supplied (type and FSC category). As for FSC-certified suppliers, the validity and scope of their certificate shall be regularly checked through the FSC online database.[5]

With regards to sourcing of Controlled Wood and reclaimed materials, reference is to be made to paragraphs 7.4 and 7.5 below. FSC-STD-40–004 also provides specific requirements for the generation and use of raw materials on-site, i.e., the case of production waste unintentionally produced and recovered for being used in the same process or sold. As a general rule, such material may be classified as the same or lower material category as the

*Table 7.1* Eligible inputs for FSC products and related output categories

| *Eligible inputs* | | *Short description* | *Output categories it can be used for* |
|---|---|---|---|
| **Improve certified % content or generate certified credits*** | **FSC 100%** | FSC-certified virgin material originating in FSC-certified forests or plantations that has not been mixed with material of another material category throughout the supply chain. Supplied with an FSC 100% claim by FSC-certified suppliers. | FSC 100%, FSC Mix |
| | **FSC Mix** | FSC-certified virgin material based on input from FSC-certified, controlled and/or reclaimed sources, and supplied with a percentage claim or credit claim by FSC-certified suppliers. | FSC Mix |
| | **FSC Recycled** | FSC-certified reclaimed material based on exclusive input from reclaimed sources, and supplied with a percentage claim or credit claim by FSC-certified suppliers. | FSC Mix, FSC Recycled |
| | **Post-consumer reclaimed materials** | Forest-based material that is reclaimed from a consumer or commercial product that has been used for its intended purpose by individuals, households or by commercial, industrial and institutional facilities in their role as end users of the product. | FSC Mix, FSC Recycled |
| **Acceptable inputs not improving certified % content or generating credits*** | **Pre-consumer reclaimed materials** | Forest-based material that is reclaimed from a process of secondary manufacture or further downstream industry, in which the material has not been intentionally produced, is unfit for end use and not capable of being re-used on-site in the same manufacturing process that generated it.<br>For paper, no distinction is made between pre- and post-consumer reclaimed materials. | FSC Mix, FSC Recycled |
| | **FSC Controlled Wood** | Virgin material supplied with an FSC Controlled Wood claim by a supplier that has been assessed by an FSC-accredited certification body for conformity with FSC Controlled Wood requirements (FSC-STD-40–005 or FSC-STD-30–010). | FSC Mix, FSC Controlled Wood |
| | **Controlled Wood material** | Virgin material supplied without an FSC claim that has been assessed to be in conformity<br>to the requirements of the standard FSC-STD-40–005. | FSC Mix, FSC Controlled Wood |

* See 7.3.3 below for further details

Source: Own elaboration. Descriptions are modified from FSC, 2016.

input from which it was derived. Material reclaimed from secondary processing may also be classified as pre-consumer reclaimed material, unless it can be re-used and incorporated back into the same process.

#### 7.3.2.4 *Material handling*

The identification of FSC-certified materials is done on the basis of the supplier's material documentation. When material is received and before it is processed, the organization shall check supplier's sales documentation to verify (FSC, 2016) (i) that the supplied material quantities and quality are in compliance with what is indicated on the sales documentation, (ii) the material category and, if applicable, the associated percentage or credit claim stated for each product item or for the total number of products and (iii) that the sales documentation includes the supplier's FSC CoC or FSC Controlled Wood codes. If these elements are missing, unclear or incomplete, the supplied materials cannot be treated as certified and shall be downgraded, unless the supplier is contacted and timely provides additional information.

FSC claims are standardized claims that reflect inputs used for a certain product group as well as the system for controlling FSC claims in use (Table 7.2). Each FSC Certificate holder is assigned a unique FSC certification code by its certification body (Figure 7.3).

If there is risk that non-eligible inputs have entered FSC product, group segregation methods – in terms of physical/temporal separation or

*Table 7.2* FSC claims, input categories and system of control

| *FSC Claim* | *Input categories* | *System for FSC claims* |
|---|---|---|
| FSC 100% | FSC 100% | Transfer system |
| FSC Mix x% | All FSC eligible inputs | Transfer system<br>Percentage system |
| FSC Mix credit | All FSC eligible inputs | Transfer system<br>Credit system |
| FSC Recycled x% | FSC Recycled, post- and pre-consumer reclaimed materials | Transfer system<br>Percentage system |
| FSC Recycled credit | FSC Recycled, post- and pre-consumer reclaimed materials | Transfer system<br>Credit system |
| FSC Controlled Wood | Controlled Wood<br>All FSC eligible inputs | Transfer system<br>Percentage system*<br>Credit system** |

* Portion of the output of a claim period that has not been sold with an FSC percentage claim
** Portion of the output volume that has not been sold as FSC Mix or FSC Recycled material, on the basis of a corresponding FSC Controlled Wood credit account (see section 7.3.3 )

Source: Own elaboration

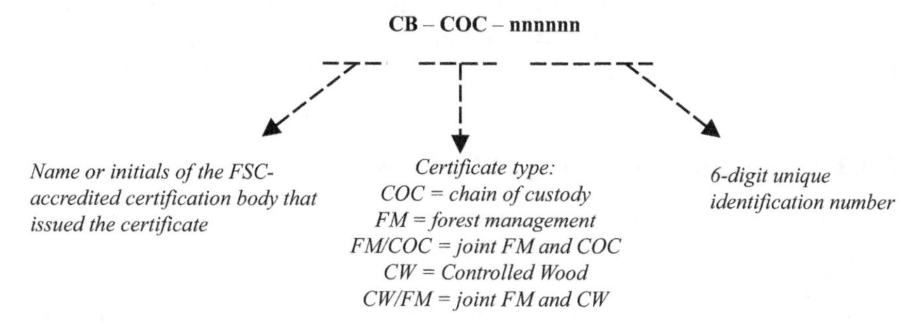


*Figure 7.3* FSC certification codes

Source: Own elaboration

identification of materials – shall be implemented. It is up to the organization to define and implement the most appropriate and effective solutions (e.g., dedicated storage areas, labels and signs, paint spots, bar-codes, etc.).

For any FSC product group or job order, the organization shall implement a material accounting system to show that at any time the quantities produced and/or sold with FSC claims are compatible with the quantities of inputs, taking into consideration their material categories, associated percentage or credit claims and conversion factor(s). Based on the accounting system in place, the organization shall prepare annual volume summaries for each product group, providing information on quantities of each material category received/used and product type produced/sold.

#### *7.3.2.5 Sales and delivery*

Sales invoices and related delivery documents issued for outputs sold as FSC certified shall include the organization's FSC CoC or FSC Controlled Wood code and a clear indication of the relevant FSC claim(s) for each product item or the total products.

#### *7.3.2.6 Compliance with timber legality legislation*

The organization shall ensure that its FSC-certified products conform to all applicable timber legality legislation. This includes that if FSC-certified products are imported/exported, procedures shall be in place to ensure conformity to all applicable trade and customs laws. Moreover, information on species (common and scientific name) and country of harvest shall be provided to direct customers and/or any FSC-certified organizations further down the supply chain. In case of products containing pre-consumer reclaimed wood to be sold in countries where timber legality legislation applies, pre-consumer material shall conform to FSC Controlled Wood requirements

(FSC-STD-40–005, see 7.4) or customers shall be informed about the presence of pre-consumer reclaimed wood in the product and supported in their Due Diligence System as required by applicable timber legality legislation.

### 7.3.3 Systems for controlling FSC claims

For each product group, the organization shall select (at least) a system for controlling FSC claims. CoC standards define three different systems:

- **Transfer system.** All outputs from a certain job order or claim period can be sold with the same claim as inputs. If inputs have different claims, the claim with the lowest FSC or post-consumer input per input volume shall be transferred to the output.
- **Percentage system.** All outputs from a certain job order or claim period can be sold with a percentage claim according to the proportion of FSC and post-consumer reclaimed inputs. If a certain threshold has been reached (i.e., 70% for FSC Mix and FSC Recycled), the total output can also carry an FSC label. The percentage claim (FSC%) shall be determined according to the following (FSC, 2016, p. 16):

  $$FSC\% = \left(QC / QT\right) \times 100$$

  where $Q_C$ is the quantity of claim-contributing inputs,[6] $Q_T$ is the total quantity of forest-based inputs. For FSC Mix and/or FSC Recycled inputs the organization shall use the percentage claim or credit claim stated on the supplier's invoice to determine the quantities of FSC and post-consumer inputs.[7] Material supplied with a credit claim shall be used by its full quantity as FSC input or post-consumer input, respectively.[8]
- **Credit system.** Only a proportion of the outputs from a certain claim period can be sold with a credit claim equivalent to the quantity of FSC and post-consumer reclaimed inputs. This proportion can also carry an FSC label. As in the previous case, FSC Mix and/or FSC Recycled inputs shall be accounted according to their percentage or credit claim. Furthermore, each input shall be converted according to the corresponding conversion factor.[9] Converted inputs can be added to a product group credit account. The output quantity sold and/or labelled as FSC Mix or FSC Recycled shall be deducted from the available FSC credit in the corresponding credit account. The credit account is never overdrawn and at any time the organization can sell material with a corresponding credit claim up to (but not exceeding) the total FSC credit available in the credit account.

Table 7.3 summarizes the main features of the three CoC systems for controlling FSC claims.

*Table 7.3* Systems for controlling FSC claims

| *System* | *Transfer system* | *Percentage system* | *Credit system* |
|---|---|---|---|
| Shall/shall not be used: | **Shall** be used for:<br>• FSC 100% product groups<br>• Trade and distribution of finished wood products and paper<br>• Trade without physical possession<br>• Food and medicinal Non-Wood Forest Products | **Shall not** be used for:<br>• FSC 100% product groups<br>• Trade and distribution of finished wood products and paper<br>• Trade without physical possession<br>• Trade and processing of Non-Timber Forest Products (NTFPs), except for bamboo and NTFPs derived from trees (e.g., cork, resin, bark) | **Shall not** be used for:<br>• FSC 100% product groups<br>• Trade and distribution of finished wood products and paper<br>• Trade without physical possession<br>• Trade and processing of NTFPs, except for bamboo and NTFPs derived from trees (e.g., cork, resin, bark)<br>• Print processes |
| Output categories: | FSC 100%, FSC Mix, FSC Recycled and FSC Controlled Wood | FSC Mix, FSC Recycled | FSC Mix, FSC Recycled |
| Output claim: | Same FSC claim as input(s)<br>If input claims are different: lower percentage or credit claim among input FSC claims | FSC percentage claim (average FSC input) | FSC credit claim |
| FSC output: | All products from a certain job order or claim period | All products from a certain job order or claim period | A proportion of products from a certain claim period |
| Claim period length: | At least the length of time to complete a batch run including receipt, storage, processing, labelling and/or sale of the output product | Up to 12 months | Up to 3 months |
| Example*: (single job order or claim period) | | | |
| Best for**: | Exclusive FSC input | High share of FSC input | Low share of FSC input<br>Continuous production processes |

* A conversion factor =1 is assumed
** This is not mandatory under FSC standards, rather it is just a comment by authors

Source: Own elaboration

### *7.3.4 Additional requirements*

Additional requirements include requirements for outsourcing of activities within the scope of a CoC certificate to both FSC-certified and non-FSC-certified contractors. Prior to outsourcing activities to a contractor, the outsourcing organization shall inform its certification body. Furthermore, an agreement or contract covering the outsourced process with each non FSC-certified contractor shall be in place and include a clause reserving the right of an FSC-accredited certification body to audit the outsourcing contractor or operation. The agreement shall also preclude sub-contracting (i.e., contractor of a contractor) as well as the use of FSC trademarks by contractors for promotion purposes. Finally, specific procedures or work instructions shall be supplied to the contractors to assure FSC-certified material is tracked, controlled and not mixed with any other material during outsourced processing.

When outsourcing involves FSC-certified contractors and outsourcing activities fall within a contractor's CoC certificate scope, contractors are subject to audits performed by the competent FSC-accredited certification body. Outsourcing arrangements are subject to a risk analysis by the certification body and sampling for on-site audit purposes.

## 7.4 Controlled Wood

Whenever an FSC-certified organization wants to mix FSC-certified and non-FSC-certified virgin material within the same product group, the non-certified proportion shall comply with FSC Controlled Wood standard. This means the material shall be verified as not being sourced from one or more of the following categories (FSC, 2017):

- a illegally harvested wood;
- b wood harvested in violation of traditional and civil rights;
- c wood harvested in forests where High Conservation Values are threatened by management activities;
- d wood harvested in forests being converted to plantations or non-forest use;
- e wood from forests in which genetically modified tress are planted.

In order to comply with FSC requirements, an organization can buy FSC Controlled Wood from suppliers holding a valid CoC certificate that includes Controlled Wood within its scope. Material will be then invoiced with an FSC Controlled Wood claim and code. As an alternative, non-certified virgin material can be purchased from non-certified suppliers. In this case, the organization shall develop, implement and maintain a documented Due Diligence System (DDS) for material supplied without an FSC claim to be used as controlled material or to be sold with the FSC Controlled Wood claim. A DDS shall be reviewed and, if necessary, revised at least annually, and consists of (i) obtaining information on material (suppliers, purchase documentation, country of harvest, information on the supply chain, etc.); (ii) risk assessment; and (iii) risk mitigation. There are several types of risk assessment, according to a well-defined hierarchy (Figure 7.4). The organization shall refer to the highest listed risk assessment that exists for the supply area, beginning with the national risk assessment (NRA). Information on approved and ongoing NRA are available via the FSC website. If no NRA is available, reference can be made to other risk assessment types, according to the hierarchy and given requirements. Company risk assessments shall be conducted according to the requirements and indicators provided by FSC Controlled Wood standards (Annex A to FSC-STD-40–005).

Risk assessments designate a risk status (low/specified risk) for the supply area with reference to each applicable risk requirement/indicator. If specified risk is identified mitigation measures shall be implemented. These measures

1. **National risk assessment (NRA)** developed according to *FSC-PRO-60-002 V3-0*
Shall be used by the organization if it exists.

2. **Centralized national risk assessment (CNRA)**
Shall be used if completed for all five Controlled Woood categories, where there is no NRA developed according to FSC-PRO-60-002 V3-0, and instead of an NRA developed according to FSC-PRO-60-002 V2-0.

3. **National risk assessment (NRA) developed according to FSC-PRO-60-002 V2-0 (“old NRAs”)**
Shall be used if there is neither an NRA developed according to FSC-PRO- 60-002 V3-0 nor a completed CNRA. It shall not be used after 31 December 2018.

4. **Company risk assessment**
May only be conducted while waiting for the delivery of the NRA or CNRA where these are scheduled.

OR

**Extended company risk assessment**
May only be conducted for unassessed risk areas where there is no FSC risk assessment available.

*Figure 7.4* The hierarchy of Controlled Wood risk assessments
Source: (FSC, 2017)

are defined within each NRA or, if no NRA exists, shall be developed by the organization itself. Any risk assessment/mitigation performed by the organization shall be assessed and approved by an FSC-accredited certification body before material is used as an input for FSC product groups. A summary of the assessment is made publicly available on the FSC online database. Material can be used as controlled only if sourced from low-risk sources or after mitigation measures are successfully implemented.

## 7.5 Reclaimed materials

When an organization wishes to use reclaimed materials as inputs for FSC product groups, reference shall be made to FSC-STD-40–007. Reclaimed materials include forest materials that would be disposed of as waste or used for energy recovery but have been collected and reclaimed as input material for re-use, recycling, re-milling in a manufacturing process or other commercial application (FSC, 2011a). The rationale behind FSC certification of reclaimed materials regards an efficient use of forest resources and, thus, the reduction of anthropic pressure on them.

The purchase of reclaimed materials for the purposes of FSC certification follows a general mechanism that is similar to the one described with regard to Controlled Wood (paragraph 7.4). An organization, indeed, can comply with FSC requirements by buying reclaimed materials from certified suppliers holding a valid CoC certificate that includes standard FSC-STD-40–007 within its scope. In this case, supplied materials will include an FSC Recycled claim and a CoC code. Otherwise, reclaimed materials can be sourced from non-certified suppliers and undergo a specific procedure for verifying sources, in order to distinguish among post-consumer and pre-consumer reclaimed materials (see Table 7.1).

According to FSC-STD-40–007, an organization shall keep a list of suppliers supplying reclaimed material. Upon receipt, all reclaimed materials shall be verified through visual inspection and classified into pre-consumer and/ or post-consumer reclaimed material. Evidence allowing classification shall be retained and can include invoices, shipping documents, pictures, material samples, statements according to official classification and grading standards,[10] quality analysis report, etc. If classification of reclaimed material is not possible at this stage, then the correspondent suppliers shall be included in the organization's Supplier Audit Program. Based on sampling, suppliers included in the program are audited by the organization itself or by an FSC-accredited certification body or other external qualified party contracted by the organization. Field audits shall take place at least annually and the minimum sampling rate shall be 80% of the square root's number of suppliers included in the Supplier Audit Program.[11] The selected sample shall alternate suppliers over time and be representative in terms of their geographical distribution, activities and/or products and size or annual production. Audits shall focus on evaluating and verifying evidence regarding the supplied material quantity, quality and compliance with FSC requirements for reclaimed material. In particular, audits shall check the existence and implementation of (i) instructions/procedures in place to control and classify the reclaimed materials; (ii) training provided to the supplier's personnel in relation to classification and control of reclaimed materials; and (iii) registers that demonstrate the origin of the materials. For each audit, a report summarizing findings shall be prepared. Reports will be counter-checked by an FSC-accredited certification body that will also perform a field assessment of a sample of suppliers of reclaimed material (see Chapter 13 for details).

## 7.6 Multisite and group certification

So far, this chapter has only considered CoC certification for single organizations trading and/or processing forest products, however, there are three types of FSC CoC certification (FSC, 2016):

- **Single CoC certification** might refer to single or multiple (i.e., two or more) sites. In the second case, only one site operates as the certificate holder, being responsible for invoicing products covered by the scope of the certificate and controlling the use of FSC trademarks (see paragraph 7.8). All sites included within the scope of the certificate shall be under a common ownership structure and direct control of the certificate holder, in an exclusive business relationship with each other for the output materials or products covered by the scope of the certificate and located within the same country. No sampling is allowed during audits performed by FSC-accredited certification bodies, i.e., all sites shall be audited at any assessment.
- **Multisite certification** includes multiple sites that either (i) are subject to common ownership (including the certificate holder) or (ii) have legal and/or contractual relationship with the certificate holder. In addition to the latter, a centrally administered and controlled management system established by the certificate holder (e.g., centralized purchases/sales, common operational procedures, same brand name, etc.) and common operational procedures (e.g., same production methods) are required.
- **Group certification** includes multiple small and independent organizations (sites) that comply with the following criteria:
    - Each organization shall qualify as small, defined by either (i) no more than 15 employees (full time equivalent) or (ii) no more than 25 employees (full time equivalent) and a maximum total annual turnover of US$1,000,000;[12]
    - All organizations shall be located in the same country as the certificate holder.

The main requirements for the three types of CoC certification are summarized and compared in Table 7.4.

Multisite and group certification types are subject to specific certification and audit requirements (see Chapter 13 for further details). The main issue consists of the distinction between a management entity, called Central Office, and single Participating Sites included within the certificate. In brief, while Participating Sites are requested to conform to all applicable FSC CoC certification requirements as summarized in this chapter, the Central Office is in charge of managing the whole multisite or group entity. This includes developing and performing a quality management system (documented procedures, training, records, etc.), establishing and implementing an Audit Program to check any applicant site before admission and to conduct at least one audit annually of each Participating Site to evaluate conformity to applicable CoC requirements, issuing Corrective Action Requested and enforcing implementation and managing the inclusion/exclusion of new sites based on evidence and results gathered from the previous audit.

Both multisite and group certificates are subject to limitations regarding the growth in number of Participating Sites. The FSC-accredited certification body will assess the capacity of the Central Office to manage the number of Participating Sites in the certificate and approve an annual growth rate up to a limit of 100% based on the number of Participating Sites at the time of the evaluation. In any case, group CoC certificates are limited to a maximum number of 500 Participating Sites.

## 7.7 Project certification

Project certification refers to renovation or once-off production of one or more objects using FSC-certified inputs. The most common use of this certification has occurred so far in the building sector, for the certification of new or renovated buildings (e.g., offices, sport facilities, etc.). Project certification may be partial or full (FSC, 2006). *Partial project certification* requires that at least some FSC-certified (100% and/or Mix) wood products are used for the project without a minimum threshold and restrictions on the sources of the remaining wood products. Only specific claims about the FSC-certified components of the project may be allowed (e.g., "The windows in this building are FSC certified"). *Full project certification* requires that at least 50% of the cost or volume of all wood products for the project are FSC certified (100% and/or Mix) and/or post-consumer reclaimed and that all of the remaining wood used for the project is pre-consumer reclaimed and/or FSC Controlled Wood. In this case, promotional claims can refer to the entire project (e.g., "FSC-certified airport").

The certificate is held by a project manager that is in charge of ensuring an FSC-accredited certification body that the project complies with all applicable requirements of FSC relevant standards. In particular, the project manager coordinates project members, i.e., all the organizations (FSC certified or not) that purchase, transform and/or install wood products for the project.

*Table 7.4* Summary comparison of single, multisite and group CoC requirements

| *Requirements* | *Single CoC* | *Multisite* | *Group* |
|---|---|---|---|
| All sites shall operate under a common ownership structure | Yes | Not necessarily<br>A legal and/or contractual relationship with the certificate holder and a centrally administered and controlled management system established by the certificate holder may suffice | No |
| Single sites can invoice FSC products independently | No, only the site acting as certificate holder is allowed to invoice FSC products | Yes | Yes |
| All sites shall be classified as small enterprises according to given criteria | No | No | Yes |
| All sites shall be located in the same country | Yes | No | Yes |
| The organization (certificate holder) shall establish a Central Office for administration and internal monitoring | No | Yes | Yes |
| The certification body can apply a given sampling method during audits | No<br>All sites included within the scope of the certificate shall be annually audited by the certification body | Yes | Yes |
| Growth in number of sites | The inclusion of new sites is allowed upon approval by the certification body | Between two consecutive audits new members may be included but growth limits established by the certification body shall be applied | Between two consecutive audits new members may be included but growth limits established by the certification body shall be applied<br>Maximum size (i.e., number of sites): 500 |

Source: Modified from (FSC, 2016)

## 7.8 Labelling

Labelling requirements are defined by FSC-STD-40–004 (Part III); however, that just indicates when a certain product is eligible to be FSC-labelled (Table 7.5), while specific rules for the use of trademarks are provided by a dedicated standard, FSC-STD-50–001. Product labelling with FSC trademarks, however, remains voluntary; when a product is eligible to carry an FSC label, it's up to each certified organization to decide whether to label it or not.

FSC trademarks – the FSC "checkmark-and-tree" logo, the initials "FSC" and the name "Forest Stewardship Council" – are registered trademarks owned by the Forest Stewardship Council A.C.; therefore, any use must be authorized and in conformity with given requirements. In order to label a product with FSC trademarks, an organization must have signed an FSC Trademark License Agreement and hold a valid CoC certificate. Moreover, labelled products shall be included in the organization's FSC product group list. Labelling shall be in line with graphic and other requirements laid down by FSC-STD-50–001 (colour, positioning, minimum size, etc.) and requires prior approval by an FSC-accredited certification body. Once an organization is certified, it receives an e-mail with credentials to log-in to an online trademark portal (label generator) and produce their own FSC labels.

Apart from labelling, FSC-certified organizations are allowed to use FSC trademarks to promote their certified products as well as their status as certified organizations outside their products (e.g., on websites, catalogues, brochures, stationery, magazines, etc.). This can be done by means of specific artwork, called FSC promotional panels, that can be

*Table 7.5* FSC labels and eligibility criteria for them

| *Type of FSC label* | *Label* | *Systems for controlling FSC claims* | *Minimum threshold* |
|---|---|---|---|
| FSC 100% | | Transfer system | 100% (exclusive use of FSC 100% inputs) |
| FSC Mix | | Transfer system<br>Percentage system<br>Credit system | 70%*<br>70%<br>Available credit |
| FSC Recycled | | Transfer system<br><br>Percentage system<br><br>Credit system | 70%* post-consumer reclaimed only for wood; for paper no threshold applies<br>70% post-consumer reclaimed only for wood; for paper no threshold applies<br>Available credit |

* When an FSC Mix (or Recycled) credit claim is transferred there is no minimum threshold

Source: Own elaboration

developed and downloaded through the above-mentioned FSC online trademark portal. Specific requirements for the use of the panel are set by FSC-STD-50–001.

In 2015, FSC launched the new "Forests for All Forever" brand, developed based on a marketing survey. The brand can be used in addition to an on-product label and/or off-product promotional panel to pass a stronger and more effective message on FSC certification.

## Notes

1 The forest gate is the place where a change in ownership of forest materials/products from a certain Forest Management Unit occurs. Depending on the case, the forest gate might coincide with forest stump, landing areas, on-site concentration yards, off-site mills or log-yards, auction houses, etc. For example, in the case of sales of standing trees, the forest gate corresponds to the forest stump, while if the transfer of ownership occurs when products (e.g., logs) are unloaded at the purchaser's facility, the forest gate corresponds to the facility itself (e.g., off-site mill).
2 FSC claims in sales documents are required in cases where subsequent customers want to use the FSC-certified products as inputs for the manufacturing of other certified products or for re-sale as FSC certified.
3 For example, if by sawing 1 $m^3$ of logs, 0.8 $m^3$ sawn wood is obtained, the conversion factor for this process is 0.8 (or 80%).
4 For example, wood species are normally not required for fibre and particle products.
5 FSC Certificate Database: https://info.fsc.org/certificate.php.
6 Input materials that count towards the determination of the FSC Mix or FSC Recycled claims for products controlled under the percentage or credit system. Eligible claim-contributing inputs are the following: FSC-certified materials (FSC 100%, FSC Mix and FSC Recycled), post-consumer reclaimed materials and pre-consumer reclaimed paper. The latter does not include other pre-consumer reclaimed materials, such as wood and cork (FSC, 2016).
7 For example, if the FSC claim on supplier's invoice reads 1 $m^3$ FSC Mix 80%, this quantity shall be accounted as 0.8 $m^3$.
8 For example, if the FSC claim on supplier's invoice reads 1 $m^3$ FSC Mix Credit, this quantity shall be accounted as 1 $m^3$.
9 For example, if an input is supplied with an FSC claim reading 1 $m^3$ FSC Mix 80% and presents a conversion factor equal to 0.8, it generates a credit of 0.64 $m^3$, i.e., 1 $m^3$ x 0.8 x 0.8.
10 For example, EN 643, The European List of Standard Grades of Recovered Paper and Board.
11 That is, where y is the number of suppliers to be audited (rounded to the upper whole number) and x the number of suppliers included in the Supplier Audit Program.
12 These are internationally valid requirements, however, nationally specific eligibility criteria can be developed by FSC National Offices based on procedure FSC-PRO-40–003.

## References

FSC (2006). FSC Chain of Custody Standard for Project Certification. FSC-STD-40–006 V1–0 EN. Forest Stewardship Council International Centre, Bonn.

FSC (2011). Policy for the Association of Organizations with FSC. FSC-POL-01–004 V2–0 EN. Forest Stewardship Council International Centre, Bonn.
FSC (2011a). Sourcing Reclaimed Material for Use in FSC Product Groups or FSC Certified Projects. FSC-STD-40–007 V2–0 EN. Forest Stewardship Council International Centre, Bonn.
FSC (2013). Evaluation of the Organization's Commitment to FSC Values and Occupational Health and Safety in the Chain of Custody. FSC-PRO-20–001 V1–1 EN. Forest Stewardship Council International Centre, Bonn.
FSC (2016). FSC Standard for Chain of Custody Certification. FSC-STD-40–004 V3–0 EN. Forest Stewardship Council International Centre, Bonn.
FSC (2017). Requirements for Sourcing FSC Controlled Wood. FSC-STD-40–005 V3–1 EN. Forest Stewardship Council International Centre, Bonn.

Chapter 8

# Forest carbon offsetting and standards

*Alex Pra and Lucio Brotto*

This chapter provides an overview of standards applicable to forest-based carbon projects. Paragraph 8.1 introduces the concept of carbon offsetting and the markets for trading of carbon offsets: compliance, voluntary and domestic markets. Paragraph 8.2 focuses on the need for rules and regulation in the voluntary carbon market, presenting its functioning and the infrastructure created to ensure quality and transparency, such as carbon standards, registries, quality programs for offset provided and carbon footprint standards. Forest-based carbon projects are described in paragraph 8.3, which also presents their applicability under the different carbon markets, the key elements (baseline, leakage permanence, additionality, co-benefits and double counting) and the methodologies. The last paragraph (paragraph 8.4) provides an excursus on forest carbon standards. Particular attention is paid to VCS, the CCB Standard, the Gold Standard, Plan Vivo, ACR and the UK Woodland Carbon Code.

## 8.1 Carbon offsetting definition and markets

Public awareness towards climate change threats has risen considerably in recent years, together with the number of businesses, organizations and individuals looking to reduce their greenhouse gas (GHG)[1] emissions and minimize their impact on global warming. "Carbon footprint" is the term used to define the total set of GHG emitted by a business, organization or individual.

Beyond the primary strategies to lower the carbon footprint, i.e., improving energy efficiency in homes or factories and changing consumption patterns by buying local products and reducing flight travel, carbon offsetting has gained prominence as a tool to compensate carbon emissions. A carbon offset (or credit) can be defined as a "credit representing the reduction, avoidance or sequestration of a metric ton of carbon dioxide or GHG equivalent"[2] (Ecosystem Marketplace, 2018). In other words, a credit for reducing the carbon footprint by paying someone else to reduce, avoid or absorb the release of a ton of carbon dioxide or GHG elsewhere. This process can happen at the worldwide level, given that global warming is not a localized issue.

In this context, markets for buying and selling carbon offsets have reached, in recent years, a substantial dimension and a high level of complexity and dynamism. Carbon markets exist under institutional regulatory schemes (compliance carbon markets) and voluntary initiatives (voluntary carbon markets). In addition, a third cross-cutting category of domestic initiatives either regulated or voluntary-based can be identified (domestic carbon markets). The next three sub-paragraphs present an overview of these markets, in terms of definition, main actors and functioning. A snapshot of the state of carbon markets can be found in Chapter 5.

### *8.1.1 Compliance markets*

Compliance markets (called also regulated or mandatory markets) are those created and regulated by mandatory national, regional or international schemes. At the global level, the biggest compliance market is created by the Kyoto Protocol of the United Nations Framework Convention on Climate Change (UNFCCC), the principal international regulation for the reduction of GHG emissions. The Kyoto Protocol entered into force in 2005. The Kyoto Protocol establishes binding targets for developed countries that have ratified the protocol (Annex B[3] countries), that consist of reducing their GHG emissions to 5.2% below their 1990 levels, through:

- reducing their own emissions;
- trading emissions allowances with countries that have allowances surplus;
- purchasing carbon credits.

In order to facilitate fulfillment of the targets, the Kyoto Protocol introduced three so-called flexible mechanisms: the Clean Development Mechanism (CDM), Joint Implementation (JI) and Emissions Trading System (ETS). The first two, CDM and JI, are non-domestic project-based mechanisms for carbon offsetting. The CDM is a mechanism whereby developed countries (Annex 1 countries[4]) can finance and trade carbon emission reduction credits from projects in developing countries (non-Annex-1 countries) to contribute meeting their own targets. In this case, the amount of emission reduction must be certified by the CDM Executive Board and the credits take the name of Certified Emission Reductions (CERs). The JI scheme works similarly to CDM, with the exception that projects can also be set up in Annex 1 countries, and in this case, credits are named Emission Reduction Units (ERUs).

The International Emissions Trade (IET) system is a market-based system that allows Annex B countries to trade their credits or allowances among themselves. In this system, countries that have fulfilled their targets can sell their credits or allowances to other countries that need to meet their Kyoto Protocol targets. In order to meet its Kyoto Protocol targets, the European Union (EU) put in place the third mechanism: its own regional scheme, the European Union Emissions Trading Scheme (EU ETS). The EU ETS establishes a cap-and-trade system with national emissions caps and limited tradable allowances, where participation is mandatory for businesses of a certain size and operating in certain sectors. Kyoto Protocol mechanisms, together with EU ETS, represent the current most active and largest carbon offset trading scheme in the world.

With the end of the Kyoto commitments in 2020, the new framework for global climate action will be defined by the Paris Agreement, adopted in Paris in December 2015 by the UNFCCC. The Paris Agreement has a profoundly different approach than the Kyoto Protocol and its new framework of rules and procedures for cooperation approaches is likely to influence carbon offsetting. An overview of the Paris Agreement's main developments related to carbon offsetting is presented in Box 8.1.

### Box 8.1 The Paris Agreement post-2020 framework for global climate action

The Paris Agreement was adopted in December 2015 in the UNFCCC 21st Conference of the Parties (COP-21) and is to come into effect and be implemented by 2020. Although the Paris Agreement and the Kyoto Protocol are two separate and independent instruments of the UNFCCC, the Paris Agreement is considered the successor of the Kyoto Protocol and it will provide a new framework for global climate action post-2020.

The Paris Agreement is fundamentally different from the Kyoto Protocol in its approach and in relation to the rules and procedures for cooperation against climate change. First of all, the agreement sets the end of the separation between developed and developing countries, and all ratifying countries are expected to contribute to the global commitment to keep the global temperatures increase below 2°C and to strive for limiting global warming to 1.5°C, and to reach a global decarbonization by the second half of this century.

All ratifying countries are expected to make and implement climate commitments, called Nationally Determined Contributions (NDCs), to contribute to the overall goal. In particular, Article 6 of the Paris Agreement ("countries can cooperate with each other to ramp up their climate change strategies and promote sustainable development") contains several decisions that could strongly influence carbon offsetting. Article 6.2 mentions that "countries can meet their NDCs by trading emission reductions ('internationally transferred mitigation outcomes') amongst each other, and they can create their own governance structures to manage the process" through use of internationally transferred mitigation outcomes towards Nationally Determined Contributions. Thus, it will be possible to derive internationally transferred mitigation outcomes from any kind of bilateral, regional or multilateral cooperation, a process not based on a centralized capped mechanism but based on bottom-up voluntary cooperation between countries (Prell, 2015). In addition, Articles 6.4 to 6.7 make arrangements for the establishment of a new mechanism that could replace CDM and JI. Article 6.4 states that "the UNFCCC will also create a centralized trading platform that countries can use to trade emissions reductions. Some are calling this the Sustainable Development Mechanism". This new mechanism will share a number of characteristics with its predecessors, but instead of delivering pure offsetting of emissions, it is intended to obtain net emission reductions, under the supervision of a body designed by the UNFCCC. Finally, Article 6.5 defines new enhanced rules for avoiding double counting: "If one country transfers an emissions reduction to another country, then it can no longer deduct those emissions from its own carbon inventory. In other words: no double counting". Thus, emission reductions resulting from the Sustainable Development Mechanism used by a country to demonstrate achievement of its NDCs shall not be used to contribute to the achievement of the host country's NDCs (Zwick, 2016), an element that may potentially increase the demand for domestic offsets.

### *8.1.2 Voluntary carbon markets*

Emission offset initiatives and trade by businesses, organizations and individuals on a voluntary basis make up the voluntary carbon market. Buyers in the voluntary carbon markets are not mandated to buy carbon offsets, as is the case with the compliance market, but are driven by a variety of considerations, such as corporate social and environmental responsibility, ethical codes, reputational and supply chain risk. In some cases, buyers even decide to invest in carbon offset from projects demonstrating strong social and environmental benefits to local communities, because they are more interested in the development aspect rather than in GHG emissions (Hamrick and Gallant, 2017a).

The voluntary carbon market is made up of unregulated transactions of carbon offsets called Verified Emission Reductions (VERs), or, in some cases, also CERs created from CDM projects, although these represent a minor part of the market volume. Given that in the voluntary market demand is created voluntarily and not by regulatory instrument, the total offset trading volumes are smaller compared to the compliance market.

The lack of rules and regulations, together with the generally lower transaction costs of projects compared to a CDM project, leads to a much greater dynamism and innovation in the voluntary carbon market in terms of project types, quality standards, methodologies and technology. Often, innovative standards and methodologies experimented with in the voluntary market are then adopted or adapted to regulatory schemes in the compliance market. However, on the negative side, the lack of common rules and regulations often leads to a lack of quality and transparency in the voluntary offset projects (see paragraph 8.2).

### *8.1.3 A cross-cutting category: domestic carbon markets*

In recent years, several governments created their own regional, national or sub-national markets for carbon offset; these are called "domestic carbon markets". The driving forces behind the creation of domestic carbon markets are the weak processes for setting new international compliance markets under the UNFCCC for the post-Kyoto world, which leaves space for bilateral and multilateral initiatives, and the buyer's preference for locally sourced carbon offsets.

The World Bank defines domestic carbon markets (Klein et al., 2015) as:

- markets that apply to organizations operating on a national scale;
- markets supported or managed by governmental organizations;
- markets with rules, standards and registries managed on a national scale;
- markets that generate carbon credits sold only at a national level;
- markets that generate carbon credits used to meet national policies (e.g., the Emissions Trading Scheme), carbon taxes, national carbon neutralization schemes or voluntary carbon market on national or local scale.

Therefore, domestic carbon markets are a cross-cutting category in the sense that they can be both under regulatory schemes, or voluntary-based.

At the regional and national scales there are examples of domestic markets created and regulated by governments. Important regulated domestic carbon initiatives at the national scale are Australia's Emissions Reduction Fund, New Zealand's ETS and South Korea's ETS; at the state/regional level, there are California's Cap-and-Trade, British Columbia's Climate Action Plan and Alberta's Offset Credit System. Even at the local level, an example is the compliance-based ETS in the neighbouring, mainly urban prefectures of Tokyo and Saitama (Tokyo-Saitama ETS).

Some of these domestic carbon market programs are on a voluntary basis, where the government organizes a carbon pricing system but it is up to businesses, organizations and individuals to participate, or not. For example, Japan has the J-Credit system, a consolidated government-managed voluntary carbon offsetting program, as well as Japan's Joint Crediting Mechanism (JCM), a government-administered program that partners with developing countries to invest in carbon offset projects. Also, Mexico has a voluntary domestic carbon trading platform called MexiCO2. The United Kingdom, in addition to participating in the EU ETS, has created the Woodland Carbon Code (WCC), a voluntary domestic offsetting program for forest-based carbon projects (Hamrick and Gallant, 2017a).

## 8.2 Quality infrastructures in voluntary carbon markets

As explained in the previous paragraph, voluntary carbon markets are made up of unregulated transactions of carbon offsets. Figure 8.1 synthetizes the structure and functioning of the voluntary carbon market. The financial flow goes from the supply side to the demand side. The supply side is represented by the economic actors able to offer carbon offset investments or quotas. These are the project owners, i.e., individuals, businesses or organizations owning or managing forests. Project owners may be supported by donors, usually non-governmental organizations (NGOs) willing to provide complementary funding for some project activities. The demand side is represented by economic actors willing to buy carbon offsets, e.g., businesses, organizations and individuals willing to compensate their emissions. These are the end-buyers, who buy the carbon offset with the intention to retire it. In between, several intermediaries and buyers may exist:

- wholesalers, which are businesses buying carbon offsets on behalf of their clients for corporate responsibility and marketing purposes (e.g., a company with the aim of offsetting the GHG emissions connected to a specific product or service in order to market it as "carbon neutral");
- retailers, which are traditionally businesses buying a portfolio of offsets that they then offer to end-buyers together with other carbon management services (i.e., advising on emission reduction strategies);
- brokers, which are business intermediaries who do not take ownership of the offset but facilitate the transaction between sellers and buyers by charging a commission fee;
- aggregators, which are businesses or organizations that buy carbon offsets and then sell in bulk, taking advantage of market prices.

Project developers are individuals or organizations responsible for developing and implementing the carbon offset projects. In some cases, the project developer can be the project owner itself, however, in most of cases they are an external entity. Project developers can be supported by consultants for carrying out specific activities, i.e., stakeholder consultation and remote sensing analysis.

The lack of common rules and regulations for the voluntary market leads to voluntary offsetting projects often being criticized for their non-additionality, that is, carbon offsets coming from projects that would have been implemented anyway, lack of social and environmental benefits for the hosting communities, lack of transparency and quality assurance through independent verification and vulnerability (see paragraph 8.4 for key issue definitions). There have been also cases of fraudulent manipulation of measurements to claim more carbon offsets from a project, sale of

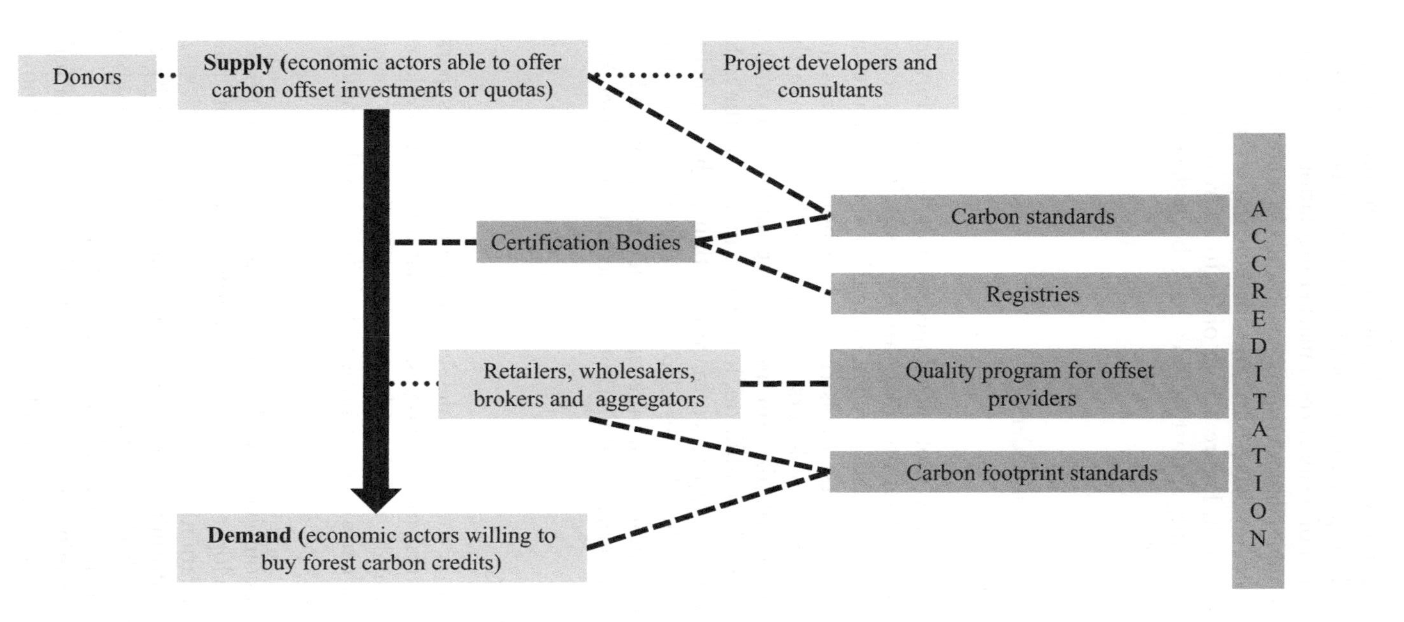


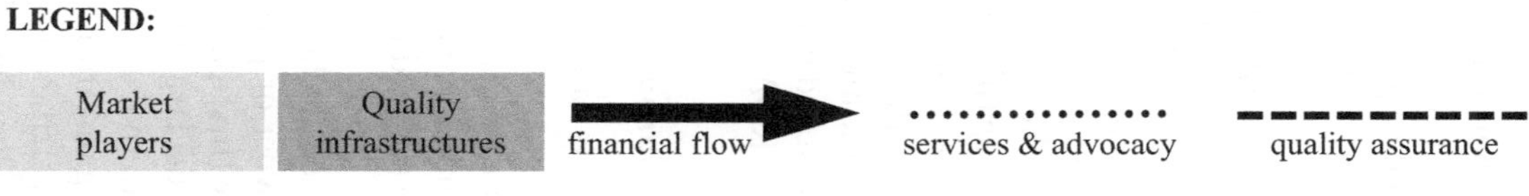


*Figure 8.1* Quality infrastructure in the voluntary carbon market

Source: Own elaboration

carbon offsets that either did not exist or belonged to someone else, false or misleading claims with respect to the social or financial benefits of an investment and exploitation of weak carbon market regulations to commit financial crimes such as money laundering or tax fraud. As a consequence, ensuring credibility and quality has gained considerable importance in the voluntary market. As shown in Figure 8.1, in order to address these challenges, several so-called quality infrastructures have been developed and put in practice in recent years:

- carbon standards;
- registries;
- quality programs for offset providers;
- carbon footprint standards.

Carbon footprint standards are tools developed mainly for end-buyers, while quality programs for offset providers are tools specifically created to ensure market quality and transparency of market intermediaries as well as end-buyers in some cases. Registries and carbon standards are tools for project owners and developers. Certification bodies have the role to check and monitor organization and project compliance with a specific certification standard. In addition, in order to guarantee the independence, impartiality and competence of certification bodies in delivering certification, accreditation bodies also assure that the certification body is monitored. An example of an independent accreditation body for carbon standards is the International Carbon Reduction and Offset Alliance (ICROA).

### *8.2.1 Carbon standards*

Standards can be defined as the set of rules, codes, protocols, norms, paradigm, measures, benchmarks, methodologies, etc., that a project developer voluntarily decides to comply with in order to ensure quality in the design and implementation of a carbon offset project. Reliability of the standard can be ensured by:

- a self-declaration of compliance with a certain standard by the project owner or project developer itself, without external assessment of validity (first-party certification);
- a declaration of conformity against a certain standard issued by an external certification body not fully independent from the assessed organization or standards (second-party certification);
- a certificate of conformity against a certain standard issued by a fully independent third-party certification body (third-party certification).

Thus, standards alone cannot ensure the quality of projects, and therefore, it is only through independent third-party certification that carbon standards can be reliably evaluated.

To improve reliability, more and more project developers are carrying out their activities in compliance with carbon standards, and the offsets are consequently certified accordingly. More than 99% of the offsets transacted in the voluntary forest carbon market are certified according to a third-party independent accredited standard. Examples of forest carbon standards can be found in paragraph 8.4.

### *8.2.2 Registries*

Registries are tools where carbon offsets are listed and tracked through their lifetimes, with the main function of avoiding double counting. Each compliance market, also at domestic scale, has its own designated registry. In the voluntary market, there are independent international registries such as the Market Environmental Registry and APX Registry.

### *8.2.3 Quality programs for offset providers*

Quality programs for offset providers are tools developed to ensure the quality and transparency of carbon offsets offered by intermediaries (typically wholesalers and retailers) to end-buyers. An example is the Quality Assurance Standard (QAS).

### *8.2.4 Carbon footprint standards*

Carbon footprint standards allow for a certified calculation of carbon emission footprint of organizations, businesses, products, projects or events (e.g., though Life Cycle Assessment (LCA) or Environmental Product Declaration (EPD) methodologies) and thus are used by actors willing to buy carbon offsets to offset their carbon footprint. An example is the Carbon Trust Standard (CTS). The main methodological standard for carbon footprint calculation is ISO/TS 14067.

## 8.3 Carbon offsetting from forest - based carbon projects

Carbon offsetting projects can be carried out in different sectors. We can distinguish five broad categories of application (Kollmuss et al., 2008):

- biological sequestration;
- renewable energy projects;
- energy efficiency projects;

- industrial gases projects;
- methane capture projects.

Biological sequestration comprises all those activities related to forestry, agriculture and other land uses. Under the UNFCCC, these activities are also defined with the acronym LULUCF (Land Use, Land Use Change and Forestry).

### 8.3.1 Types of forest-based carbon projects

Forestry has a significant impact on climate change. On the one hand, forests act as a carbon sink given their ability to sequester and store carbon dioxide from the atmosphere; on the other hand, as a carbon source, emissions from forest degradation and deforestation are estimated to account for a 20% of anthropogenic GHG emissions.

We can distinguish three types of projects for climate change mitigation from forest-based activities:

- Those aiming to avoid emissions thought the conservation of forests, i.e., avoiding deforestation or avoiding degradation of existing forests (i.e., Reducing Emissions from Deforestation and Forest Degradation (REDD)) (Figure 8.2).
- Those aiming to increase carbon storage though planting trees or converting non-forested land to forested land, i.e., afforestation and reforestation (A/R) projects. Afforestation stands for planting trees in areas that have never been forested in recent history, while reforestation means the restoration of forest cover in areas previously cleared or burned, naturally or for agriculture (Figure 8.3).
- Those aiming to increase carbon stocks by planting trees without actually resulting in the creation of a forest, i.e. urban forestry, agroforestry systems, forest densification or enrichment, forest restoration.
- Those aiming to increase carbon storage with Improved Forest Management (IFM), i.e., climate-friendly cultivation techniques, less intense logging, etc.

Beyond sequestration and storage of carbon dioxide by the atmosphere, forest-based projects have a unique capacity to provide a wide range of

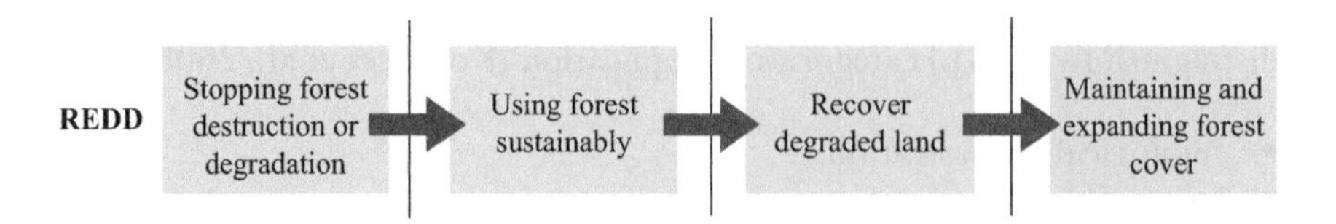


*Figure 8.2* Reducing Emissions from Deforestation and Forest Degradation example

Source: Own elaboration

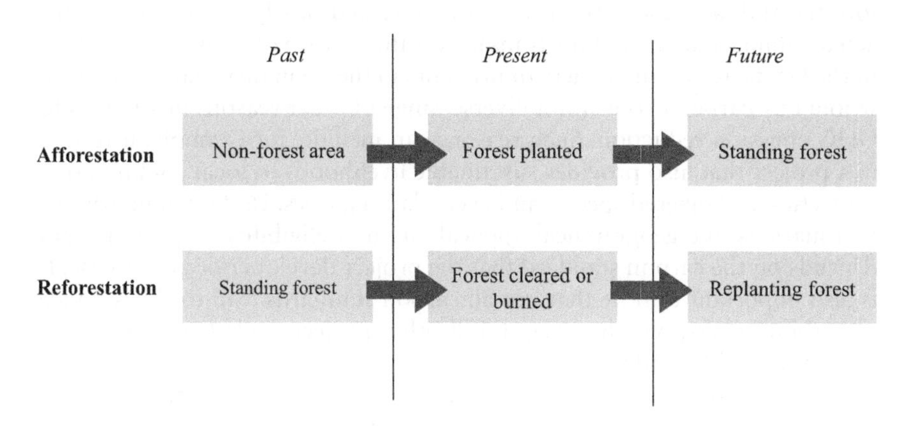


*Figure 8.3* Afforestation and reforestation examples

Source: Own elaboration

ecosystem services (i.e., water flow regulation, habitats for biodiversity, tourism and recreation, etc.) and livelihoods for local communities through sustainable forest management (i.e., construction timber, fruits and other Non-Wood Forest Products, biomass for energy, etc.).

The drawback of the sector for carbon projects developers is the complexity to reliably calculate the carbon sequestration potential and to establish adequate baselines. A forest's carbon sequestration depends on many factors (age, growth rate, climate, soil, management, etc.). In addition, this complexity is growing due to the impact of climate change, which determines increased temperatures, altered precipitation regimes and changes in disturbances regimes, and can change the forest's ability to sequester and store carbon.

### 8.3.2 Applicability of forest-based carbon projects under compliance, voluntary and domestic markets

Forest-based carbon projects are almost entirely excluded from the compliance markets. In the market mechanism under the Kyoto Protocol, forest-based projects play a marginal role. Only afforestation and reforestation activities are identified as eligible for the CDM, but they make up less than 1% of CDM registered projects (see Chapter 5). The EU ETS dost not allow regulated businesses to buy forest-based carbon offsets to meet their obligations. This is due to concerns that offsets from forest-based projects do not result in permanent carbon sequestration (i.e., risk of forest fire or unplanned forest loss) and the complexity of reliable carbon accounting.

Most of the carbon offsets from forest-based carbon projects are sold on the voluntary market. Hamrick and Gallant (2017a) reported that in 2016, forestry and land use offset projects represented nearly 27% of the transacted volume and 46% of the transacted value in the voluntary market. One of the key factors is the variety of investors in the voluntary market, some of whom buy carbon offsets for a diverse range of other reasons in addition to GHG emission reduction. Such reasons can include, for example, investing in a project that also provides sustainable livelihoods to local communities, protecting endangered species and even planting trees. Under voluntary carbon markets, the geographical applicability and eligibility of project types depends on the carbon standard that the project developer decides to use. It is also important to note that not all carbon standards lead to the issuance of carbon offsets, but they may fulfill other project goals (e.g., social and environmental benefits).

More recently, several domestic market initiatives, both compliance-based and voluntary, have begun to accept offsets from forest-based carbon projects. For example, California's Cap-and-Trade program and Australia's ERF began in 2013. The United Kingdom created the Woodland Carbon Code (WCC) in 2011 as a voluntary domestic offsetting program to include forest-based carbon projects (in this case specific to woodland creation). Other governments in Europe are moving in the same direction, e.g., France's Voluntary Carbon Land Certification program, specifically targeting forestry and land use carbon projects at the domestic level. Another example is the voluntary-based Forest Carbon Offset Scheme of South Korea, which is specific for forest-based carbon projects.

### 8.3.3 Key elements in forest-based carbon projects

When developing forest-based carbon projects, project developers have to address several issues carefully, i.e., permanence of the investment, additionality of the project, the avoidance of risks related to leakage, the social and environmental impacts of the project and the risks of double counting or double claiming. All these key elements are introduced in the following sections.

#### *8.3.3.1 Baseline*

The baseline represents the business-as-usual scenario; that is, it describes the future conditions and outcomes that would be expected if no project activity were to take place. Project GHG benefits are therefore calculated against a baseline scenario. It is important that the baseline is accurately estimated and verifiable and the project GHG benefits calculated upon the baseline, as not to occur in overestimations. The baseline can be estimated using project-based or performance-based methodological approaches, as is the case with additionality. Examples of hypothetical baseline scenarios and project impacts are presented in Figure 8.4 (Ebeling and Olander, 2011).

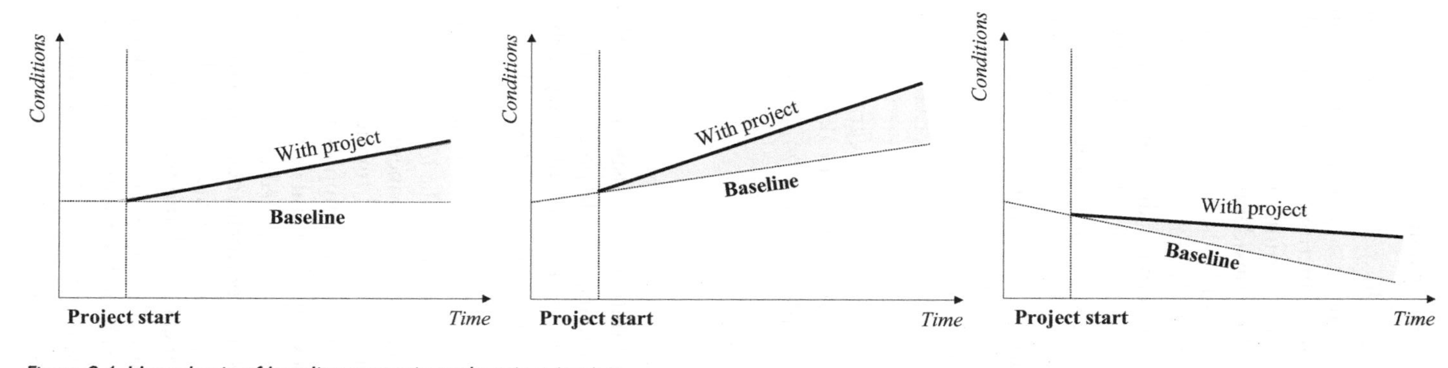


*Figure 8.4* Hypothesis of baseline scenario and project impact

Source: Own elaboration based on Ebeling and Olander (2011)

#### *8.3.3.2 Project boundaries and leakage*

Project boundaries refer to physical, legal and organizational boundaries of the project, which is necessary to define accurately in order to be able to calculate emissions reduction. In particular, the issue of leakage is of crucial importance when defining project boundaries in forest-based carbon projects and careful consideration of leakage risk is a requirement in all forest carbon standards. Leakage is defined as the project's unanticipated increase of GHG emissions outside the project boundaries, whether in the form of activity shifting or market leakage, e.g., if a project that aims to reforest a pastureland drives local farmers to clear forest elsewhere to open new pastures. If leakage assessment procedures are relatively well-established in forest carbon standards (in particular for afforestation and reforestation projects), on the other hand, it can be very complex to assess leakage risk due to the difficulty of collecting the data needed to meet accounting requirements. Appropriate activities for leakage risk mitigation include creating alternative sources for fuelwood and timber and improving efficiency in their usage, creating alternative employment opportunities, conducting integral zoning and development of plans that preclude activity shifting. Leakage that cannot be avoided in the development of the project must be quantified and debited from overall project benefits.

#### *8.3.3.3 Additionality*

A fundamental and often controversial issue in all carbon offset projects is additionality, which is an indispensable precondition for all projects both under compliance market schemes and voluntary carbon standards. Additionality can be defined as the proof that the carbon offset project would have not been implemented anyway. In other words, only those projects that would not have been implemented anyway should be counted for carbon offsetting. This principle supports the idea that it is essential for carbon offset buyers to know that their investment is decisive for the project to happen and it is not just an additional financial benefit for project developers. Additionality, it is often very complex to determine in practice, and carbon standards have developed several methodologies in order to address this challenge. These can be distinguished in two broad categories: project-based additionality tests and performance standards. Project-based additionality tests assess each project on a case-by-case basis; the most common types of tests include financial, legal and technological barrier assessments. For example, a financial additionality test examines whether the project activities could have been carried out without the financial incentives from carbon offset (if yes, the project is not additional). While project-based additionality tests assess projects on a case-by-case basis, performance standards assess technologies or processes that generate GHG emission reductions. These can be benchmark approaches or positive technology lists. The first consists of establishing a threshold against which projects are measured, while the second consists of providing a list of all technologies and processes that will be considered additional based on the context. Performance standards shift part of the costs and administrative burden to a standard organization, making additionality testing less costly for project developers.

#### *8.3.3.4 Permanence*

Permanence refers to the duration and stability of a carbon stock within its management and disturbance regimes. Therefore, it comprises the risk of re-emission of the sequestered carbon in the atmosphere, i.e., through clearing, burning or dieback. The issue of permanence is a distinctive property of forestry and land use projects and its management and mitigation is specifically addressed in forest carbon standards, i.e., through the issuing of temporary offsets, or buffer offset reserves, where a certain amount of offset is not sold and is kept aside in case of problems.

#### *8.3.3.5 Co-benefits*

Co-benefits include all social and environmental benefits that an offset project delivers to local communities. In this sense, forest-based projects are particularly attractive because of the capacity of forests to provide a wide range of ecosystem services and livelihoods for local communities. Co-benefits are becoming increasingly important for the value of the project and its attractiveness for buyers who are more and more concerned about corporate responsibility.

#### *8.3.3.6 Double counting*

Double counting refers to the claim, selling, issuance or double monetization of a carbon offset, which can happen inadvertently or intentionally. Thus, double counting can take different forms and can occur both in compliance and voluntary markets. Four forms of double counting can be identified (Nett and Wolters, 2017):

- Double selling, when a carbon offset is sold more than once to different buyers.
- Double issuance, when the same carbon offset is accounted and credited twice (or more) under two different standards or registries.
- Double claim, when two actors claim the environmental benefit of the exact same carbon offset. For example, if one carbon offset is used by

a company in the voluntary market, but also accounted in a national inventory and claimed by the host government.
- Double monetization, in the case of project sectors that are listed in the national GHG inventory, when a carbon offset is monetized as a GHG allowance (e.g. under Kyoto Protocol or EU ETS) as well as a VER in the voluntary market.

Cases like double selling and double issuance are related to accounting and registration problems that can be addressed with the use of international and centralized registries where carbon offsets are registered and tracked. The cases of double claiming and double monetization are more critical with regard to the environmental integrity of a carbon offset.

### *8.3.4 Methodologies for forest-based carbon projects*

A methodology for forest-based carbon projects summarizes the assumptions and mathematic formula to use to calculate baseline, permanence, additionality, leakage, social and environmental impacts and double counting for a specific project.

Methodologies vary in their suitability for different types of forest-based carbon activities (for example, for reforestation, rather than for avoided deforestation or agroforestry activities) and location. A project developer willing to apply for carbon standards (see paragraph 8.4) needs to use a methodology that is approved by the standard. For example, CDM provides a set of consolidated methodologies for A/R projects. CDM provides four methodologies, two for large-scale A/R CDM project activities and two for small-scale:

- AR-AM0014: afforestation and reforestation of degraded mangrove habitats (*large-scale*);
- AR-ACM0003: afforestation and reforestation of lands except wetlands (*large-scale*);
- AR-AMS0003: afforestation and reforestation project activities implemented on wetlands (*small-scale*);
- AR-AMS0007: afforestation and reforestation project activities implemented on lands other than wetlands (*small-scale*).

For other types of forest-based projects, a number of methodologies have been developed under voluntary standards. Depending on the standard requirements, the project developer can also revise pre-existing methodologies or develop completely new ones.

## 8.4 Forest carbon standards

As explained in paragraph 8.2, to address the challenges of assuring credibility and quality in the voluntary carbon market, many offset quality standards have been developed in recent years. At the moment, approximately a dozen third-party certified standards applicable to forest-based carbon projects can be identified in the voluntary carbon market:

- Verified Carbon Standard (VCS);
- Climate, Community and Biodiversity (CCB) Standard;
- Plan Vivo system;
- the Gold Standard;
- American Carbon Registry (ACR).

In terms of the market share, according to Hamrick and Gallant (2017b), in 2016, VCS certified the majority of forestry and land use projects (82%), and 73% of these were certified in conjunction with the CCB Standard. The next largest share of the market was made up of ACR certified offsets (5%), mostly located in the United States. The Gold Standard and Plan Vivo, both with a strong emphasis on co-benefits, accounted respectively for 4% and 2% of the market volume. Other widespread standards included the Climate Action Reserve (CAR) and Social Carbon Standard.

All these standards differ on approved methodologies, eligible project categories and location, price and other project development and implementation requirements. In addition, it should be noted that not all standards lead to the issuance of carbon offsets, but might fulfill different aims, i.e., certifying co-benefits or ensuring sustainable forest management. In order to better understand the different standards, we can distinguish them as follows:

- *methodological standards* that mostly focus on defining specific and reliable methodologies for accounting carbon stock in forests and generate carbon offsets to be traded on the market (e.g., VSC, the Gold Standard, Plan Vivo system);
- *standards for co-benefits* that emphasize social and environmental benefits from the implementation of carbon projects, with special attention on biodiversity and/or local communities and beneficiaries, including investors (e.g., CCB Standard, Social Carbon Standard).

In addition, there are standards for sustainable forest management, like FSC®, which can be adopted as a combined certification with specific carbon standards. For example, VCS standards for carbon accounting acknowledge FSC and CCB standards as valuable references for enlarging the set of benefits from carbon projects. In compliance markets, forest-based carbon projects are developed under regulated schemes such as the CDM, and more recently, also national and sub-national markets have designed new standards specific to domestic needs for voluntary use, i.e., the UK-WCC.

Table 8.1 provides an overall comparison of the of the main standards used in forest-based carbon projects, indicating the type of standard (if it requires a third-party verification), scope (methodological standard, standard

*Table 8.1* Summary of forest carbon standards

| *Standard* | *Third-party verification* | *Scope** | *Eligibility* | | | *Methodologies for carbon quantification* | *Additionality assessment* |
|---|---|---|---|---|---|---|---|
| | | | *Project type*** | *Location* | *Size restrictions* | | |
| CDM | Yes | M | A/R | Non-Annex-I countries | No | Yes | Yes, project-base[d] additionality test[s] and performance standards |
| VCS | Yes | M | A/R, IFM, RED, URB | Any | No | Yes | Yes |
| CCB | Yes | CB | All | Any | No | No, but project must demonstrate a net reduction of GHG emission | Yes, project-base[d] additionality test[s] |
| The Gold Standard | Yes | M | A/R | Any | No | Yes | Yes, through A/R CDM tool OR Positive List |
| Plan Vivo | Yes | M | All | Developing countries | No | Yes | Yes, project-base[d] additionality test[s] |
| American Carbon Registry | Yes | M | A/R, IFM, REDD | Any (but certain methodologies specific for US) | No | Yes | Yes, project-base[d] additionality test[s] OR approved performance standards |
| UK-WCC | Yes | M | A/R | UK | No | Yes | Yes, project-base[d] additionality test[s] |
| Climate Action Reserve | Yes | M | A/R, IFM, REDD, URB, WET | US | No | Yes | Yes, through lega[l] requirement test and performance test |
| Social Carbon | Yes | CB | All | Any | No | No | No |
| FSC | Yes | SFM | A/R, IFM, REDD | Any | No | No | No |

* M = Methodological Standard, CB = Standard for Co-Benefits, SFM = Sustainable Forest Management Standard
** A/R = Afforestation and Reforestation, IFM = Improved Forest Management, REDD = Reducing Emissions fro[m] project types mentioned in paragraph 8.3

Source: Own elaboration

| ermanence equirements | Co-benefits | Leakage avoidance | Credits/offsets issued | Registry |
|---|---|---|---|---|
| es, temporary mission credits | Yes, sustainability criteria developed by host countries | Yes, specific methods are developed under each baseline methodology | Certified Emissions Reductions (CERs) | CDM registry |
| es, credit buffer eserve | No, but facilitate the combination with CCB Standard | Yes, specific to each project type | Verified Carbon Units (VCUs) | Markit Environmental Registry and APX |
| lo, but permanence isk must be ddressed in the roject design phase | Yes, specific criteria for biodiversity and stakeholders engagement | Yes | No carbon credits issued | Internal |
| es, credit buffer eserve (20%) | Yes, Sustainable Development Assessment tool | Yes, based on national or international default values | Planned Emissions Reduction Units (PERs) | Markit Environmental Registry |
| es, several nechanisms (i.e., nsold reserve of redits, long-term ale agreement, long- erm landownership) | Yes, detailed requirements and metrics | Yes, using an integrated approach to planning | Plan Vivo Certificates (also *ex-ante*) | Markit Environmental Registry |
| es, risk assessment nd monitoring | No | Yes, specific methodologies | Emissions Reductions Tons (ERTs) | APX |
| es, several nechanisms (risk ssessment, carbon uffer reserve, emonstrated long- erm management ntentions) | Yes, stakeholder consultation and Environmental Impact Assessment | Yes, risk assessment | Woodland Carbon Units | Markit Environmental Registry |
| es, credit buffer eserve | No | Yes, specific to each project type | Climate Reserve Tonnes (CRTs) | Internal |
| lo | Yes, Sustainable Livelihood Approach | No | No carbon credits issued | Markit Environmental Registry |
| lo | Yes, criteria and indicators defined in the standard | No | No, carbon credits issued | Internal |

forestation and Forest Degradation, URB = Urban Forestry, WET = Wetland Restoration, All = includes all forestry for co-benefits, others), project eligibility criteria (type, location, size), methodologies for carbon quantification, additionality assessment, permanence requirements, co-benefits requirements, leakage avoidance methods, the type of carbon offset issued and the registry system it adopts. Please note that standards are quickly evolving, thus it is recommended to always reference the standards' official normative documents, which can be generally found on the official websites.

The next section provides an overview of the main standard used for forest-based carbon projects. Information described in these sections comes mainly from standards websites, which are reported in the chapter notes.

### *8.4.1 Verified Carbon Standard (VCS)*[5]

The Verified Carbon Standard (VCS)[6] is currently the most used standard for voluntary carbon offset projects. VCS was funded by the Climate Group, the International Emissions Trading Association and the World Business Council for Sustainable Development, and the first version of the Standard was released in 2006.

The current version of the Standard is v3.7, released in June 2017 (VCS, 2017a). The standard sets the requirements for developing projects, programs and methodologies. In addition, the VCS Standard is supported by other documents that provide requirements specific to Agriculture, Forestry and Other Land Use (AFOLU) projects (VCS, 2017b), jurisdictional programs and nested REDD+ projects (VCS, 2017c). VCS category AFOLU includes the following types of projects:

- afforestation and reforestation;
- Improved Forest Management (IFM), i.e., conventional to reduced impact logging, converting logged to protected forest, extending rotation age, conversion of low-productive forests to productive forests;
- Agricultural Land Management (ALM), i.e., improved cropland management, improved grassland management and cropland to grassland conversion;
- Reducing Emissions from Deforestation and Forest Degradation;
- Avoided Conversion of Grasslands and Shrublands (ACoGS);
- Wetlands Restoration and Conservation (WRC).

Another document provided by VCS that is useful for project developers is the VCS Program Guide (VCS, 2017d), which describes rules and requirements governing the VCS Program, i.e., registration process, registry system, methodology approval process and accreditation requirements for validation/verification bodies. The VCS Standard has no location or size restrictions. It does not monitor social and environmental benefits of the projects; simply, project developers have to demonstrate no negative social and environmental impacts caused by the project. It issues Verified Carbon

Units (VCUs) and it is connected to Markit Environmental Registry and APX Registry.

### *8.4.2 Climate, Community and Biodiversity (CCB) Standard*[7]

The Climate, Community and Biodiversity (CCB) Standard was created by the Climate, Community and Biodiversity Alliance (CCBA), a unique partnership of leading international NGOs (Conservation International, the Nature Conservancy, CARE, Rainforest Alliance and others) and research institutes, and since November 2014 it is managed by the VCS.

The CCB Standard is a project design standard focusing on strengthening the design and assuring additional environmental and social criteria and benefits to carbon offset projects. The CCB Standard is not a carbon accounting standard, so it does not verify carbon emission reduction or issue offsets. Thus, in order to sell offsets from CCB verified projects tagged with the CCB label, project developers must use them in combination with a carbon accounting standard (e.g., VCS). CCB Standard certified projects can both be developed as a CDM or for selling VERs in the voluntary market. As of May 2017, 102 CCB projects have been validated in 32 countries and another 11 were undergoing validation (VCS, 2017d).

The first edition of the CCB Standard was released in 2005. The current reference version is CCB Standard v3.1 released in June 2017 (CCB, 2017a) after an update of the previous version 3.0 of December 2013. The Standard defines all specific requirements for developing and monitoring projects, focusing exclusively on forest and agriculture projects, such as:

- primary or secondary forest conservation;
- afforestation and reforestation;
- agroforestry plantations;
- densification and enrichment planting;
- introduction of new cultivation, harvesting or processing practices.

Eligible projects have no location or size restrictions. Although it is not a carbon accounting standard, the CCB Standard still requires the project to demonstrate a net reduction of GHG emission, quantified according to approved methodologies (e.g., IPCC). CCB provides project developers with two other useful guidance documents. The first is the CCB Program Rules (CCB, 2017b), which presents all the rules and requirements governing the CCB Program. The second is the CCB Program Definitions (CCB, 2017c), which provides the definitions for terms used in the CCB Program documents.

### *8.4.3 The Gold Standard*[8]

The Gold Standard was established in 2003 by WWF and other international NGOs as a high quality standard for officially registered CDM projects.

It has recently enlarged its scope and is one of the leading standards for voluntary-based climate and development projects.

In 2018, it launched a new standard, the Gold Standard for the Global Goals (Gold Standard, 2018a), to certify a range of UN Sustainable Development Goals (SDG) impacts in addition to carbon offset. Among forest and land use projects, only afforestation and reforestation projects are eligible under the Gold Standard. The specific requirements for afforestation and reforestation projects are funded in the Gold Standard for the Global Goals Land Use and Forests Activity Requirements, published in March 2018 (Gold Standard, 2018b). The Gold Standard also partners with FSC to promote sustainable forest management and it facilitates the process of obtaining a parallel dual certification.

### 8.4.4 *Plan Vivo system*[9]

The Plan Vivo system is an offset standard developed for smallholders and community-based projects using a "payments for ecosystem services" approach. The standard emphasizes participatory approaches in the design and implementation of the project, with a specific focus on poverty reduction and livelihood development, restoration of degraded ecosystems and biodiversity conservation and adaptation of communities to climate change.

The Plan Vivo system originates from a research project in southern Mexico in 1994 carried out by the Edinburgh Centre for Carbon Management (ECCM) in partnership with El Colegio de la Frontera Sur (ECOSUR), the University of Edinburgh and other local organizations. Today, Plan Vivo is managed by the Plan Vivo Foundation.

The Plan Vivo Standard for Community Payment for Ecosystem Services Programs (Plan Vivo, 2013) was released in 2013 and accepts the following project types:

- afforestation and reforestation;
- agroforestry;
- avoided deforestation;
- forest conservation and restoration.

In terms of location, Plan Vivo projects are based only in developing countries and there is no size limitation. At present it counts three projects in Latin America, nine in Africa and eight in Asia and Pacific. Plan Vivo defines specific eligibility requirements also for producers and project coordinators. Producers must be "small-scale farmers, land-users or forest dwellers with recognized land tenure and user rights [. . .] and must be organized in democratically controlled cooperatives, association or community-based organizations". Project coordinators must be an "established legal entity [. . .] with a strong in-country presence and experience to work effectively with smallholders and local communities".

The Plan Vivo Foundation reviews and registers projects according to the Plan Vivo Standard and issues Plan Vivo Certificates following the submission and third-party verification of project's annual reports. Both *ex-ante* and *ex-post* offsets are approved under the Plan Vivo Standard, depending on the project and activity; however, in order to ensure sufficient start-up funds for project participants, they are usually issued on an *ex-ante* basis. Plan Vivo Certificates are recorded and tracked through the Markit Environmental Registry. Plan Vivo provides two extra documents that the project developer should consult: the Plan Vivo Procedures Manual (Plan Vivo, 2017), which provides an overview of the Plan Vivo project cycle, and the Plan Vivo Socio-economic Assessment Manual (Plan Vivo, 2016), which gives further information on socio-economic aspects.

### 8.4.5 American Carbon Registry[10]

The American Carbon Registry (ACR) was founded in 1996 and represents the first private voluntary offset program in the world. It is part of the non-profit organization Winrock International. It operates in both the voluntary carbon market and as an approved Offset Project Registry (OPR) for the California Cap-and-Trade program. Forest-based project types accepted by ACR include:

- afforestation and reforestation;
- Improved Forest Management;
- Reduced Emissions from Deforestation and Degradation (REDD);
- wetland restoration.

The ACR Standard v5.0 was released in February 2018 (ACR, 2018) and details ACR's requirements and specifications for the quantification, monitoring, reporting, verification and registration of project-based GHG emissions reductions and removals. It is applicable at the global level (although most of the projects under ACR certification are based in the USA), and issues Emission Reduction Tons (ERTs). In addition, ACR has approved a Nested REDD+ Standard (ACR, 2012).

### 8.4.6 UK Woodland Carbon Code[11]

The Woodland Carbon Code is a specific standard for voluntary-based woodland creation projects in the UK. It has a specific focus on woodland creation on soils that are not organic, defined as "the direct, human-induced conversion to woodland of land that has not been under tree cover for at least 25 years [. . .] established by planting, direct seeding or natural regeneration". The Woodland Carbon Code is in its second version released in March 2018 (UK Forestry Commission, 2018).

## Notes

1 Gases and other warming agents that contribute to climate change, i.e., carbon dioxide ($CO_2$), methane ($CH_4$), nitrous oxide ($N_2O$), hydrofluorocarbons (HFCs), perfluorocarbons (PFCs) and sulphur hexafluoride ($SF_6$).
2 "A measure of the global warming potential of a particular GHG compared to that of carbon dioxide. One unit of a gas with a $CO_2$ rating of 21, for example, would have the global warming effect of 21 units of carbon dioxide emissions (over a time frame of 100 years)" (Ecosystem Marketplace, 2018).
3 The 39 emissions-capped industrialized countries and economies in transition listed in Annex B of the Kyoto Protocol.
4 "Annex 1" refers to the UNFCCC and "Annex B" to the Kyoto Protocol. Note that Belarus and Turkey are listed in Annex 1 but not Annex B; and Croatia, Liechtenstein, Monaco and Slovenia are listed in Annex B but not Annex 1. Source: www.cdmcapacity.org/glossary.html
5 http://verra.org/project/vcs-program/
6 Previously called the Voluntary Carbon Standard (VCS)
7 http://verra.org/project/ccb-program/
8 www.goldstandard.org
9 www.planvivo.org
10 https://americancarbonregistry.org
11 www.forestry.gov.uk/carboncode

## References

ACR (2012). Nested REDD+ Standard. Requirements for Registration of REDD+ Projects Nested within a Jurisdictional Accounting Framework. Version 1.0 – October 2012. Available at: https://americancarbonregistry.org/carbon-accounting/standards-methodologies/american-carbon-registry-nested-redd-standard/acr-nested-redd-standard-2017.pdf

ACR (2018). The American Carbon Registry Standard: Requirements and Specifications for the Quantification, Monitoring, Reporting, Verification, and Registration of Project-Based GHG Emissions Reductions and Removals. Version 5.0 – February 2018. Available at: https://americancarbonregistry.org/carbon-accounting/standards-methodologies/american-carbon-registry-standard/acr-standard-v5-0-february-2018.pdf

CCB (2017a). Climate Community and Biodiversity Standards. Third Edition. 21 June 2017, v3.1. Available at: http://verra.org/wp-content/uploads/2017/12/CCB-Standards-v3.1_ENG.pdf

CCB (2017b). CCB Program Rules. Third Edition. 21 June 2017, v3.1. Available at: http://verra.org/wp-content/uploads/2017/12/CCB-Program-Rules-v3.1.pdf

CCB (2017c). CCB Program Definitions. Third Edition. 21 June 2017, v3.0. Available at: http://verra.org/wp-content/uploads/2017/06/CCB_Program_Definitions_v3.0.pdf

Ecosystem Marketplace (2018). Glossary. Ecosystem Marketplace, Washington. Available at: www.ecosystemmarketplace.com/glossary

Gold Standard (2018a). Gold Standard for the Global Goals Principles & Requirements. Version 1.1 – March 2018. Available at: https://globalgoals.goldstandard.org/wp-content/uploads/2018/02/100-GS4GG-Principles-Requirements-v1.1.pdf

Gold Standard (2018b). Gold Standard for the Global Goals Land-Use & Forests Activity Requirements. Version 1.1 – March 2018. Available at: https://globalgoals.goldstandard.org/wp-content/uploads/2017/07/200-GS4GG-Land-Use-Forests-Activity-Requirements-v1.1.pdf

Hamrick, K., Gallant, M. (2017a). *Fertile Ground: State of the Forest Carbon Finance 2017*. Ecosystem Marketplace, Washington, DC.

Hamrick, K., Gallant, M. (2017b). *Unlocking Potential: State of the Voluntary Carbon Markets 2017*. Ecosystem Marketplace, Washington.

Klein, N., Borkent, B., Gilbert, A., Esser, L., Broekhoff, D., Tornek, R. (2015). Options to Use Existing International Offset Programs ina a Domestic Context. Technical Note 10, August 2015, Washington, DC, United States of America.

Kollmuss, A., Zink, H., Polycarp, C. (2008). *Making Sense of the Voluntary Carbon Market: A Comparison of Carbon Offset Standards*. WWF Germany, Berlin.

Nett, K., Wolters, S. (2017). *Leveraging Domestic Offset Projects for a Climate-Neutral World: Regulatory Conditions and Options*. German Emission Trading Authority at the German Environment Institute, Berlin.

Olander, J., Ebeling, J. (2011). Building Forest Carbon Projects: Step-by-Step Overview and Guide. In *Building Forest Carbon Projects*, Johannes Ebeling and Jacob Olander (eds.). Forest Trends, Washington, DC.

Plan Vivo (2013). Plan Vivo Standard for Community Payment for Ecosystem Services Programmes. Available at: www.planvivo.org/docs/Plan-Vivo-Standard.pdf

Plan Vivo (2016). Plan Vivo Socio-Economic Manual Integrating Livelihood and Participatory Approaches into the Design, Development and Monitoring of Plan Vivo Projects. Updated August 2016. Available at: www.planvivo.org/docs/Socio-economic-Manual.pdf

Plan Vivo (2017). Plan Vivo Procedures Manual for the Registration and Oversight of Plan Vivo Projects and Issuance of Plan Vivo Certificates. Version – May 2017. Available at: www.planvivo.org/docs/Procedures-Manual.pdf

Prell, C. (2015). The Paris Agreement on Climate Change: A Practical Guide. December 15 2015. Available at: www.crowell.com/NewsEvents/AlertsNewsletters/all/The-Paris-Agreement-on-Climate-Change-A-Practical-Guide

UK Forestry Commission (2018). Woodland Carbon Code. Requirements for Voluntary Carbon Sequestration Projects. Version 2.0 – March 2018. ISBN 978-0-85538-843-0. Available at: www.forestry.gov.uk/pdf/WWC_V2.0_08March2018.pdf/$FILE/WWC_V2.0_08March2018.pdf

VCS (2017a). VCS Standard Version 3. Requirements Document. June 21, 2017, v3.7. Available at: http://database.verra.org/sites/vcs.benfredaconsulting.com/files/VCS_Standard_v3.7.pdf

VCS (2017b). VCS Agriculture, Forestry and Other Land Uses (AFOLU) Requirements. VCS Version 3. Requirements Document. June 21, 2017, v3.6. Available at: http://database.verra.org/sites/vcs.benfredaconsulting.com/files/AFOLU_Requirements_v3.6.pdf

VCS (2017c). VCS Jurisdictional and Nested REDD+ (JNR) Requirements. VCS Version 3. Requirements Document. 21 June 2017, v3.4. Available at: http://database.verra.org/sites/vcs.benfredaconsulting.com/files/Jurisdictional_and_Nested_REDD%2B_Requirements_v3.4.pdf

VCS (2017d). VCS Program Guide. VCS Version 3. Requirements Document. June 21, 2017, v3.7. Available at: http://database.verra.org/sites/vcs.benfredaconsulting.com/files/VCS_Program_Guide_v3.7.pdf

Zwick, S. (2016). The Road from Paris: Green Lights, Speed Bumps, and the Future of Carbon Markets. Available at: www.ecosystemmarketplace.com/articles/green-lights-and-speed-bumps-on-road-to-markets-under-paris-agreement/

Chapter 9

# Woody biomass certification standards

*Ondřej Tarabus and Nicola Andrighetto*

This chapter provides an overview of woody biomass certification standards with a specific focus on the Sustainable Biomass Program. After providing a global overview of the wood energy sector (paragraph 9.1), the current sustainability requirements in different countries (paragraph 9.2) as well as existing certification schemes (paragraph 9.3) are described. Paragraph 9.4 presents, in detail, the Sustainable Biomass Program and paragraph 9.5 introduces the ENplus® certification as an example of quality certification for woody biomass. Finally, some conclusions are provided (paragraph 9.6).

## 9.1 Wood energy sector: definition and renewable energy share

The term "bioenergy" means any type of energy produced from organic matter. Recently, important factors, such as public incentives, the decreasing cost of renewable energy generation technologies and the uncertain prices of fossil fuels, have favoured a global increase in bioenergy demand (World Energy Council, 2016). In 2016, bioenergy covered almost 90% of renewable heat use and 8% of renewable power consumption. As a result, in 2016, bioenergy was the largest renewable energy source in primary energy consumption (REN21, 2016). Solid biomass, including wood, charcoal, agricultural and forestry waste, is the most important single source of renewable energy, providing about 6% of the world's total primary energy supply annually (IRENA, 2016). The domestic use of solid biomass for energy is predominant in Africa, but it also has a significant role in industrialized countries (FAO, 2014). Indeed, more than 80 million people in Europe and around 8 million people in North America use woody biomass as their main heating source (FAO, 2014); the production of electricity from solid biomass is mainly concentrated in Europe and the USA. Indeed, more than 70% of all solid biomass consumed to produce electricity is utilized in the EU and USA (World Energy Council, 2016).

However, this rapid increase in demand for solid biomass for energy production has created worry in civil society, especially at in Europe. Firstly, there is concern surrounding whether the future woody biomass demand for energy of the EU countries represents a threat for European forests. That said, one of the most cited EU wood studies (Mantau et al., 2010) suggests that, if woody biomass were actually to play the role ascribed in every national renewable energy action plan, in 2030, the demand for forest biomass would increase by 73% compared to 2010. This would lead to the consequent risk of forest biomass shortages within the European Union. This strong utilization of wood for energy could also result in competition with other traditional sectors, such as the construction and paper and pulp industry. As a result, if energy prices were to rise, the effect of this would be to divert the price of timber for industrial purposes (European Commission, 2014).

Furthermore, doubts have been raised about whether the strong and uncontrolled increase of utilization of woody biomass for energy could compromise the main benefits that justify most public incentives (Fritsche and Iriarte, 2014). The carbon neutrality of woody biomass, for example, could be threatened by long-distance transportation. In relation to this, EU imports (both intra- and extra-regional) of woody biomass for energy doubled in terms of weight and tripled in terms of value between 2000 and 2014. In particular, in the same period, EU imports from non-EU countries grew sevenfold in weight and tenfold in value.[1]

Another issue relates to the role of plantations for satisfying the EU solid biomass demand. In the near future, EU imports of woody biomass originating from planted forests in less developed countries is expected to increase. Planted forests are rapidly increasing and often require intense management (use of fertilizers and pesticides), resulting in potentially negative environmental impacts as well as stimulating competition with other land uses and potential conversion of natural forests (de Schutter and Giljum, 2014).

These numerous concerns have led a growing number of stakeholders, such as NGOs, to demand that the procurement of raw material meets a set of minimum sustainability requirements (Fritsche and Iriarte, 2014). Indeed, liquid biofuels used in the transport sector are regulated by the Renewable Energy Directive (RED), but no binding sustainability requirements on solid biomass exist. At the moment, there is a draft document of a new directive (RED II) being discussed; the final document is expected to be ready by the end of 2018. In order to cover this gap, some European countries have implemented sustainability requirements for solid biomass, which has resulted in the need for certification systems that can cover these requirements and provide a system to verify that the biomass is sourced responsibly.

## 9.2 Current solid biomass sustainability requirements

The main importing countries for solid biomass (UK, Denmark, the Netherlands and Belgium) have adopted, to some extent, some sustainability requirements for biomass. The requirements differ significantly country by country, which creates trade barriers and brings additional administrative costs for demonstrating compliance. This has led to increased demand for a comprehensive scheme to cover all the requirements.

### 9.2.1 UK sustainability requirements

The UK government has established sustainability requirements for biomass feedstock used for the production of heat, in power plants and for co-firing in existing coal power plants. The Department for Business, Energy and Industrial Strategy (BEIS) and the Department of Transport introduced four support mechanisms: the Renewables Obligation (RO), the Renewable Heat Incentive (RHI), the Contracts for Difference (CfD) and the Renewable Transport Fuel Obligation (RTFO). Each of these includes sustainability requirements for solid biomass. Among these initiatives, the RO is the UK government's main support mechanism to incentivize deployment of large-scale renewable electricity generation. The aims of the sustainability requirements for solid biomass under the RO are to deliver real GHG savings whilst assuring that solid biomass is produced in a way that does not give rise to deforestation or degradation of habitats or loss of biodiversity.

The UK has developed comprehensive criteria, and these criteria were adopted as the baseline for the comparison of the national sustainability requirements. Although the evaluation was based on the UK requirements, it was also expanded to include additional requirements and criteria that are not included in the UK system, such as cascading and carbon debt criteria.

### 9.2.2 Danish sustainability requirements

There are no mandatory sustainability requirements for solid biomass used in the Danish energy sector, but a voluntary industry agreement (IA) was established by the Danish District Heating Association and the Danish Energy Association in 2014. The IA aims to support the use of solid biomass (chips and wood pellets) for energy production in Denmark, and attempts to comply with the Danish framework for sustainability in terms of the environment, health and safety and climate. The combined heat and power producers are themselves responsible; the producers document and satisfy requirements for sustainability through a third party.

The requirements for sustainable biomass were developed based on the most comprehensive biomass sustainability legislation that existed at the time, namely the UK Sustainability Criteria for Solid Biomass, and do not apply for agricultural feedstock.

### 9.2.3 Dutch sustainability requirements

The Netherlands was one of the first countries to implement sustainable criteria for solid biomass. Criteria was defined by the Cramer Commission and included, among others, carbon stocks, GHG emissions, biodiversity, impact on soil, water, air or competition with food production. After the criteria were approved, voluntary certification schemes were developed (e.g., Better Biomass).

In 2013, the Netherlands Enterprise Agency (RVO) published the Energy Accord for Sustainable Growth, stating that biomass used for co-firing and heat production must meet sustainability criteria. As part of the Dutch 2013 Energy Accord, the Sustainable Energy Production Incentive Scheme (SDE+) has been introduced including sustainability criteria set in legislation based on Better Biomass requirements, and additionally covering carbon debt, ILUC and SFM requirements. These requirements do not apply to secondary feedstock (produced from industrial waste such as sawdust or by-products).

### *9.2.4 Belgium sustainability requirements*

In 2002, Belgium introduced mechanisms to promote the usage of renewable sources for electricity production through a quota system based on obligations, tradable certificates and minimum prices, as well as sustainable certification and subsidies for investment in and utilization of renewable electricity.

The trade of certificates is subject to federal legislation, while the quota obligations are defined in regional regulations. Electricity suppliers need to show evidence that they have supplied a certain quota of renewable energy determined by three regions – Flanders, Wallonia and Brussels-Capital (which have almost the same requirements so that they are integrated as one region for further investigation) – to their final consumers. The quota systems do not include sustainability requirements for various types of renewable energy, but they require evidence of sustainable forest management for forest biomass, including certifying, or at least providing evidence thereof, the type of raw materials as well as the energy and $CO_2$ balance of the supply chain to an accredited inspection, including proof of compliance with responsible management of the forests, controlled impact on the environment and enforcement of legislation. In Flanders, additional requirements relating to the cascading use of biomass have been established.

## 9.3 Existing certification schemes

### *9.3.1 Sustainable Biomass Program*

The Sustainable Biomass Program (SBP) is a certification system designed for woody biomass, mostly in the form of wood pellets and woodchips, used in industrial, large-scale energy production. SBP was established in 2013 as a not-for-profit company, owned and fully funded by its members, all of which use woody biomass for large-scale energy production, that is, heat and power.

The main consumers of biomass reacted to the fact that there was no single certification scheme that would cover all the requirements, and at the same time one that would cover a significant part of the sourcing area. At the moment, SBP is transforming from a membership to a multi-stakeholder organization.

SBP is in line with the UK, Belgium and Danish legislative requirements and is currently undergoing the approval process of the Dutch system (where it might apply only to biomass produced from secondary feedstock). The verification of compliance with SBP standards is done by accredited certification bodies (international accreditation by ASI). The biomass can always be traced back to the producer, because each transfer is registered in the Data Transfer System (DTS), which is a blockchain system for tracking the material and the claim used by all the certificate holders. At the moment, there are around 150 certificates issued spread over 18 countries.

### 9.3.2 Better Biomass

The Better Biomass certificate is used by organizations who wish to demonstrate that the biomass they produce, process, trade and/or use meets well-established global sustainability criteria as well as chain-of-custody requirements. These have been defined by a multi-stakeholder working group, managed by the Netherlands Standardization Institute (NEN), and have been published in the standards NTA 8080–1 and NTA 8080–2

There are two types of Better Biomass certification: for biomass products that fall outside the scope of RED, i.e., apply to solid and gaseous biomass for energy purposes other than transport and to bio-based products; or for biomass products that fall within the scope of RED, i.e., apply to biomass products that are intended for biofuels (fuel for transport) or bioliquids (liquid fuel for other energy purposes).

The first certification type is relevant for solid biomass. This scheme is mostly relevant for the Dutch market and does not fulfill all the requirements of the other countries where some legislation is in place. Only certification bodies that have entered into a license agreement with NEN are allowed to perform such conformity assessments; the accreditation is also managed by NEN. At the moment, there are around 100 certificates, issued mostly in the Netherlands and Belgium.

### 9.3.3 ISCC

ISCC is an independent multi-stakeholder organization providing a globally applicable certification system for the sustainability of raw materials and products. ISCC is a multi-feedstock system that can be used to certify all types of biomass, including agricultural or forestry raw materials, waste and residues. The ISCC certification system is applicable to entire supply chains and for different sectors and markets, including bioenergy (biofuels and bioliquids), food, feed and bio-based products, e.g., biochemical or bioplastics.

In addition, there is a certification type called ISCC Plus under the ISCC system, which might be used for solid biomass as well. So far, there are not many certificates issued for solid biomass and the system mostly focuses on biofuels and compliance with existing RED. However, it is expected that when RED II is approved by the EU, this certification system will be more active also in solid biomass certification.

ISCC and the American National Standards Institute (ANSI) have signed an agreement to cooperate in an independent third-party program to accredit certification bodies that will be conducting ISCC certification. At the moment, there are over 17.000 certificates issued in more than 100 countries, mostly for agricultural commodities or waste materials such as used cooking oil.

## 9.4 Sustainable Biomass Program

SBP is a leading certification system used in solid biomass. It was established as a reaction to the low amount of FSC®/PEFC or SFI-certified material on the market with the aim of filling a gap in the regulatory framework governing the use of woody biomass for energy production that currently prevents many forest owners from participating in the biomass supply chain, and uniquely it provides a carbon accounting framework as required by many regulators. Thus, SBP encourages using existing Forest Management Certification arrangements where possible, and is open to collaboration and cooperation where this delivers equal or better assurance more efficiently. SBP's vision is an economically, environmentally and socially sustainable woody biomass supply chain that contributes to a low carbon economy.

SBP recognizes fully the credibility of existing and well-proven forest certification systems, for example, the Forest Stewardship Council (FSC) and the Program for Endorsement of Forest Certification (PEFC) schemes, and does not wish to compete with, or replicate, them. However, there is limited uptake of certification in some key forest areas and the aforementioned schemes do not yet cover all the key requirements of biomass users (such as cascading or carbon accounting). Therefore, SBP is working to develop solutions, short and long-term, to address these issues and is in discussion with both FSC and PEFC to determine how these challenges might be overcome.

To date, SBP has developed a certification system to provide assurance that woody biomass is sourced from legal and sustainable sources, allowing companies in the biomass sector to demonstrate compliance with regulatory requirements. The SBP certification system is designed as a clear statement of principles, standards and processes necessary to demonstrate such compliance. Wherever possible, use is made of the FSC and PEFC standards and processes already applied to other forest product streams. Further refinement and strengthening of these SBP standards will follow as necessary.

SBP, unlike FSC or PEFC, is a risk-based certification system that starts on the biomass producer level; this biomass producer is responsible for verification of all feedstock entering into the certified production. There are two possible streams of certified feedstock to production. Either the material is already FSC/PEFC or SFI certified, in which case the material is automatically considered as SBP compliant, or the material needs to be included in the so-called Supply Base Evolution (SBE), which is a risk-based system for verification of sustainability of the feedstock.

The first step of this SBE is to produce risk analyses for the sourcing area. In some countries, Regional Risk Assessments (RRA) are already produced by different working groups from the regions. The lack of RRA can be considered as one of the weak points of the system as the quality of the risk assessment of the biomass producer is often low, and the organizations do not have proper resources to develop a good quality risk assessment. In any case, the risk assessment needs to provide a description of risk and risk designation for each of the 38 SBP sustainability requirements.

When risk assessment is completed, the biomass producer needs to implement mitigation measures for the indicator identified with specific risk. This system provides an effective tool that focuses on high-risk issues in forestry, while no resources are used as criteria which, by default, are low risk and well-managed in the sourcing area.

The mitigation measures need to be implemented at supplier level and, in most cases, need to go down through the supply chain to the Forest Management Unit (FMU) level. Often, this includes expert training of the suppliers and sub-suppliers, field verification before harvesting, interviews with the forest workers, developing of special databases to monitor the High Conservation Values or avoiding of sourcing some type of material. This whole system (including the risk assessment) needs to be described in detail in the document call (Supply Base Report) and shared with stakeholders, who have one month to provide comments. Often, the stakeholder consultation also includes face-to-face meetings or workshops.

When third-party audits reveal that the system is well implemented, and the certificate is issued, this material can be sold as SBP-compliant biomass. Together with this claim, the biomass producer needs to develop two documents detailing the feedstock and the energy used for sourcing the feedstock, production of biomass and transportation to the customer. All this data (including the sold volume of biomass) are, at the moment of the sale, inserted into the DTS and need to be accepted by the customer.

## 9.5 Quality matters: the case of ENplus

ENplus® is a quality certification scheme for wood pellets introduced in 2010 thanks to the initiative of the German Pellet Trade Association (DEPV). The

ENplus certification scheme covers the whole pellet production chain, from the production to the store, in order to ensure the quality of products to the end users. It is based on the ISO standard 17225–2 and its final goal is the designation of different pellet quality classes. In detail, the following three classes are defined by ENplus:

- ENplus A1;
- ENplus A2;
- ENplus B.

Each class can differ for specific pellet properties, such as the moisture and ash contents and mechanical durability. Compared to the standard, ENplus contains more restrictive limits for some characteristics, especially for mechanical durability.

ENplus certification includes control of the quality characteristics of the product, as laid down in ISO 17225 standards, in each step of production cycle. In addition to the quality, ENplus certification includes an identification system to facilitate the traceability of pellet sources as well as sustainability, criteria like the requirement that the pellet producer state the amount of $CO_2$eq emissions per metric ton produced. The independence of the certification system will be guaranteed through the involvement of accredited certification organizations according to EN 45011. The inspection and tested bodies have to be accredited, respectively, according to EN ISO 17020 or EN ISO 17025.

Nowadays, the ownership of the ENplus trademark remains with the European Biomass Association (AEBIOM), which hosts the European Pellet Council (EPC). The right to award the license to use the ENplus brand to qualifying companies is passed by AEBIOM to national pellet associations that have been accepted as national licensers. In 2017, more than 9.2 million tons of pellets were produced as ENplus certified, representing 67% of the pellets used in the European heating sector (EPC, 2017). According to the ENplus website,[2] more than 200 pellet producers are ENplus certified (most of them are in Germany and Austria).

## 9.6 Conclusions

The situation surrounding certification of solid biomass is unstable and comes with a number of challenges. Due to the lack of EU-wide legislation, it is difficult for the energy producers to provide assurance that the sustainability requirements are fulfilled. The most-used scheme is SBP, however, we can expect that when RED II is approved by the EU, other biofuel schemes will aim to increase their scope for this type of material. This could be a challenging process, because the existing requirements only focus on agriculture and there are many forestry-specific requirements.

## Notes

1 Results of our analysis of data included in UN COMTRADE database (https://comtrade.un.org/). Last access on 3 April 2018.
2 ENPlus. www.enplus-pellets.eu/

## References

de Schutter, L., Giljum, S. (2014). *A Calculation of the EU Bioenergy Land Footprint: Discussion Paper on Land Use Related to EU Bioenergy Targets for 2020 and an Outlook for 2030*. Institute for the Environment and Regional Development, Vienna University of Economics and Business, Vienna, Austria.

EPC (2017). *ENplus® Certification Statistical Overview 2017*. European Pellet Council, Brussels.

European Commission (2014). State of Play on the Sustainability of Solid and Gaseous Biomass Used for Electricity, Heating and Cooling in the EU. Commission Staff Working Document. Bruxelles, Belgium.

FAO (2014). Agriculture, Forestry and Other Land Use ì. Emissions by Sources and Removals by Sinks. Working Paper Series. ESS/14–02. FAO Statistics Division, Rome.

Fritsche, U.R. and Iriarte, L. (2014). Sustainability Criteria and Indicators for the Bio-Based Economy in Europe: State of Discussion and Way Forward. *Energies*, 7(11), 6825–6836, doi:10.3390/en7116825.

IRENA (2016). *Renewable Energy Benefits: Measuring the Economics*. International Renewable Energy Agency, Abu Dhabi, United Arab Emirates.

Mantau, U., Saal, U., Prins, K., Steierer, F., Lindner, M., Verkerk, H., Eggers, J., Leek, N., Oldenburg, J., Asikainen, A., Anttila, P. (2010). EUwood: Real Potential for Changes in Growth and Use of EU Forests. Final Report. Hamburg.

REN21 (2016). Renewables 2016. Global Status Report. Renewable Energy Policy Network. United Nations Environment Programme (UNEP). Paris.

World Energy Council (2016). *World Energy Resources 2016*. World Energy Council, London.

Chapter 10

# Certification standards applicable to Non-Wood Forest Products

*Giulia Corradini, Enrico Vidale and Davide Pettenella*

Non-Wood Forest Products (NWFPs), such as greeneries, mushrooms, berries, nuts, resins, essential oils, litter, medicinal plants and other products sourced from the forest play an important role in many rural economies as a source of food and income. This role is reinforced by the fact that, all over the world, NWFP harvesting and consumption is associated with local traditions and indigenous knowledge (Shackleton et al., 2011). Commercialization is one of the oldest economic activities associated with the use of forest resources. Nowadays, commercialization is considered an activity that should be promoted as a strategic economic tool for conserving forest ecosystems and contributing to the livelihoods of people that depend on forests. However, the high demand for certain species, associated with the conditions of poverty of the people often involved in NWFP harvesting, can lead to a depletion of many species and their habitats, especially in the case of medicinal and aromatic plants (Lange, 1998). Harvesting limits, legal constrains and other command-and-control systems are usually implemented by governments in different countries to preserve species from overharvesting. In recent years, scholars, civil society and some industries have been developing NWFP certification as a tool for supporting sustainable value chains and for influencing customer behaviour toward the consumption of NWFPs and sustainable harvesting. This chapter[1] illustrates the rationale for certification of NWFPs (paragraph 10.1), introduces essential terms and definitions (paragraph 10.2) and illustrates the standards and certification schemes most relevant for NWFPs (paragraph 10.3). Finally, some conclusions are provided (paragraph 10.4).

## 10.1 The rationale behind NWFP certification

Several studies show that opportunities exist to promote the sustainable harvesting, management, trade and use of NWFPs through certification (Shanley et al., 2002; Walter, 2002; Vantomme and Walter, 2003; Burgener and Walter, 2007; Shanley et al., 2008). The benefits generated by the adoption of certification may be several: the social sphere can be improved with the control and monitoring of harvesting rights and empowerment of local producers; the economic domain may be enhanced with the creation of premium prices for the value chain actors and especially for the pickers that respect the sustainable harvesting quantities, and at the same time guarantee the entire traceability of the product along the value chain; and finally, certification can support also the ecological sustainability of the habitats where NWFPs grow, through accounting, monitoring and control of the producers during their harvesting activities in the forest. However, the process of NWFP certification is often a not simple issue. As Shanley et al. (2002) and Burgener and Walter (2007) noted, these products are a more difficult group of products to certify than timber, due to an array of factors, including their diverse and peculiar nature and social and ecological complexity. Basic legal factors such as unsafe harvesting rights can limit the applicability of certification to NWFPs from the beginning (Pierce et al., 2003). Economic barriers can hinder the process as well. This may happen because harvesting in the wild often requires high labour inputs for low values and for this reason NWFPs suffer from diseconomies of scales (Pierce et al., 2008). In addition, the production of many NWFPs is also strongly affected by seasonality, which creates discontinuity. Moreover, NWFPs are often traded on small and local scales and trade systems are not efficiently structured. Ecological and technical challenges for certification exist as well. In particular, for some species, the definition of the sustainable harvesting rate represents a difficult assessment (Walter, 2006). Finally, there is a very broad range of end uses of NWFPs, comprising food and food additives, cosmetics, pharmaceuticals components and handcrafts. Despite the challenges revealed by these studies, in many cases, the problems have been overcome and today many examples of standards and certification schemes that can be applied to NWFPs exist in the market.

## 10.2 Non-Wood Forest Products: terms and definitions

In the late 1980s, Non-Timber Forest Products (NTFPs) were defined as "all biological materials other than timber which are extracted from forest for human use". A decade later, FAO introduced a new term to label all the products generated from the forest not made of wood, Non-Wood Forest Products, defined as "all goods of biological origin other than wood, derived from forests, other wooded land and trees outside forests" (FAO, 1999). Hence, an NWFP may be gathered from the wild or produced in forest plantations, agroforestry schemes and from trees outside forests. While the term "Non-Wood Forest Product" is relatively well known among foresters, end users have a minimal understanding of it. The same happens among producers and retailers, who have rarely used NWFP as an acronym for their transaction and marketing campaigns. Indeed, NWFPs has been commonly substituted with alternative terms like "wild products", "wild harvested products", "natural (or forest) products", "wild crops" and "plants and

*Table 10.1* Terms and definitions for NWFPs adopted in different standards

| *Term* | *Definition* | *Type* | *Source* |
|---|---|---|---|
| Non-Timber Forest Products | All products other than timber derived from the Forest Management Unit. | FS | (FSC, 2014) |
| | All forest products except timber, including other materials obtained from trees such as resins and leaves, as well as any other plant and animal products. | FS | (NEPCon, 2014) |
| | Non-Timber Forest Products include game animals, fur-bearing animals, nuts and seeds, berries, mushrooms, oils, foliage, medicinal plants, peat and fuelwood, forage and Christmas trees. | FS | (CSA, 2013) |
| | Products derived from forests other than roundwood or woodchips. Examples include, but are not limited to, seeds, fruits, nuts, honey, maple syrup and mushrooms. | FS | (SFI, 2015) |
| | Economic resources other than timber products acquired from forest, forest understory or other land use, following sustainable management principles. | FS | (CFCC, 2014) |
| Non-wood products | Forest products other than wood (e.g., honey, water, wildflowers). | FS | (AS, 2007) |
| Plants growing naturally (organic plants) | Organic plants harvested by methods so as not to interfere in preserving the ecosystem in collection areas. | OS | (MAFF, 2012) |
| Wild collection | Products collected from the wild (e.g., medicinal and aromatic plants, gums and resins, wild fruits, nuts and seeds, mushrooms). | PS | (FairWild Foundation, 2010) |
| | The collected plants grow naturally in an area, which has not been treated with prohibited inputs (according to the respective organic regulation) for at least 3 years. The collected plants must grow and regenerate naturally without any agricultural measures. | OS | (IMO/SIPPO, 2005) |
| | The collection of edible plants or parts thereof, growing naturally in natural areas or forests, where the only human interference consists of the harvest (collecting) of the products. | OS | (BioLand, 2013) |
| Wild crop | Any plant or portion of a plant that is collected or harvested from a site that is not maintained under cultivation or other agricultural management. | OS | (USDA, 2011) |

(*Continued*)

*Table 10.1* (Continued)

| *Term* | *Definition* | *Type* | *Source* |
|---|---|---|---|
| Wild grown products | "Wild grown products" are defined as products that have grown without or with low influence of the operator gathering the products. | OS | (Naturland, 2014) |
| Wild harvested products | Wild harvested products shall only be derived from a sustainable growing environment. Products shall not be harvested at a rate that exceeds the sustainable yield of the ecosystem, or threatens the existence of plant, fungal or animal species, including those not directly exploited. | OS | (IFOAM, 2012) |
| Wild harvesting | The harvesting of plants, plant products and fungi from the wild (but not animals). It has also been called "wild crafting". | OS | (SA, 2014) |
| Wild plants | Edible plants, mushrooms and parts thereof, which grow naturally in forests and on farmland and are not cultivated using agricultural methods. | OS | (Bio Suisse, 2014) |
| Wild species | Organisms captive or living in the wild that have not been subject to breeding to alter them from their native state. | OS | (UEBT, 2012) |

Note: FS = Sustainable Forest Management Standard; PS = Sustainable Plant Management Standard; OS = Organic Products Standards. Source: (ITC International Trade Center, 2007) modified and updated.

plant products from collection", considered to be more easily understandable by the customers. An NWFP market analysis highlighted that most consumers recognize the single species and the related characteristics in the market, rather than a category of products, and consumers identify NWFPs as "wild" products (Kilchling et al., 2009). In certification standards, there is no convergence to a common terminology, where we can find either "Non-Wood Forest Products/Non-Timber Forest Products" or "wild collection", as illustrated in Table 10.1. Some definitions focus on the "action of harvesting" regardless of the land of origin, others focus both on biotic and abiotic products generated from the forest, while others focus only on plant products or by-products of forest management.

## 10.3 Standards and certification schemes applicable to NWFPs

A wide range of certification systems can be applied to NWFPs, and they can be differentiated according to the focus they have, which can be ascribed to the tree main dimensions of sustainability: environmental friendliness, economic viability and social equity. A certification scheme rarely addresses all three dimensions. However, many schemes do not cover only one issue but cover, to different degrees, several areas. Therefore, overlaps and potential synergies between the different certification schemes exist. The following paragraph provides an overview of the certification types of major interests for NWFPs. Each certification type is then assessed on whether it directly targets NWFPs and whether it contains ecological specifications for collection.

### *10.3.1 Sustainable Forest Management Standards*

Forest certification started after the Rio Earth Summit, mainly to drive the forest managers to achieve sustainable timber extraction, especially in tropical forests and plantations, while assessing the impact on forest exploitation through a set of principles, criteria and indicators (see Chapter 5). Forest certification refers to two processes, forest management (FM) certification and Chain of Custody (CoC) certification. FM certification is a process that verifies that the area of forest/plantation is being managed according to the standard, while CoC certification tracks forest products from the certified forest to the point of sale. Today, more than 50 forest certification standards exist, with national, regional or global scope. The two largest certification schemes are the Forest Stewardship Council (FSC®) and the Program for the Endorsement of the Forest Certification (PEFC). Both FSC and PEFC cover NWFP (also called Non-Timber Forest Product or NTFP) production and Chain of Custody.

FSC was the first global forest certification program to be established, in 1993. A few years later, the FSC Board approved a policy to allow certification and labelling of NWFPs. Chicle-gum from Mexico was the first FSC-certified NWFP in June 1999. Since then, many NWFPs have been certified all around the world, either with Forest Management Certification and Chain of Custody certification, like cork in Portugal, Spain, Oregon and Italy, maple syrup in the USA, pine resin in Belarus, pine nuts in Portugal, essential oils in Nepal, UK and Brazil and mushrooms in Poland. Some innovative products made of NWFPs have been certified as well, such as the first neoprene-free wetsuits made of natural rubber. FSC developed a system in which each FSC-endorsed organization could develop an NWFP standard, rather than formulating a unique, central standard. In practice, national/regional FSC offices may develop a national standard that includes NWFPs in the scope, as has happened recently in Italy (FSC, 2017b). Another option is when an FSC-endorsed organization can create an NWFP addendum, to be enclosed in the general standard, as occurred in Russia, where a certification body developed an addendum that covers honey, dogrose fruits, mushrooms, truffles, willow bark and willow rod (NEPCon, 2017). Each of these standards/addenda includes more or less restrictive ecological specifications, such as the need to monitor indicators for NWFPs, which could include the impact of harvest, growth rates, loss of vigour or decline and recruitment. The certification body NEPCon recently developed an NWFP addendum applicable on a global scale, where national or product-specific NWFP standards have not been developed. It states that population size of a species, structure of the population, harvest rates and growth and regeneration rates have to be recorded and monitored though specific indicators for the different NWFP types, like plant exudates, vegetative structures (apical buds, bark, roots, leaves), reproductive structures (fruits, seeds) and for all the other NWFP categories (NEPCon, 2014). Box 10.1 summarizes the approach of FSC to NWFPs.

**Box 10.1 Sustainable Forest Management Certification (the FSC example)**

- *Specificity to NWFPs*: the standards specifically target NWFPs (named NTFPs);
- *Ecological specifications*: included;
- *Scale*: national/regional standards and addenda to standards are developed on a case-by-case basis; recently, NEPCon developed an addendum that is globally applicable for several NWFPs.

### *10.3.2 Wild collection standards*

The certification for sustainable wild collection was emphasized by scholars, though it has had a limited diffusion around the world until recently (Lange, 1998). This is due to the complex and costly parameters a company needs to provide in order to measure the level of sustainable harvesting. The most significant example of sustainable wild collection certification is FairWild.[2] The FairWild Foundation standard and certification system, born in 2008, is based both on ecological and social aspects. The ecological part is based on the International Standard for Sustainable Wild Collection of Medicinal and Aromatic Plants. The standard defines guidelines and provides tools for harvesters, producers and other stakeholders for the creation of a sustainable resource management system based on the Good Agricultural and Collection Practices. To be certified, plants and fungi that grow naturally should be collected in a way that (i) plant populations do not decrease, (ii) the species survive in the long-term, (iii) their surroundings are not damaged and (iv) no other plants or animals are disturbed. Probably because FairWild certification requires a species-specific endorsement, in March 2017 relatively few species were certified under FairWild – only 17 (*Achillea millefolium*, *Adansonia digitate*, *Glycyrrhiza glabra*, *Glycyrrhiza uralensis*, *Juniperus communis*, *Malus sylvestris*, *Rosa canina (with R. rubiginosa and R. villosa)*, *Rubus idaeus*, *Rubus fruticosus*, *Sambucus nigra*, *Taraxacum officinale*, *Terminalia bellirica*, *Terminalia chebula*, *Tilia cordata*, *Tilia platyphyllos*, *Tilia tomentosa (syn. Tilia argentea)*, *Urtica dioica*). Only 10 companies have applied for the FairWild certification (FairWild Foundation, 2017). Box 10.2 summarizes the approach of FairWild to NWFPs.

**Box 10.2 Certification of wild products (the FairWild example)**

- *Specificity to NWFPs*: does not directly target NWFPs as a category, rather the sustainable collection of wild products, on a case-by-case basis;
- *Ecological specifications*: included;
- *Scale*: global, but it requires individual approval of each product.

### *10.3.3 Organic product standards*

Organic agriculture is

> a production system that sustains the health of soils, ecosystems and people. It relies on ecological processes, biodiversity and cycles adapted to local conditions, rather than the use of inputs with adverse effects. Organic agriculture combines tradition, innovation and science to benefit the shared environment and promote fair relationships and a good quality of life for all involved.
>
> (IFOAM, 2008)

An increasing number of consumers, in some countries more than others, have changed their purchasing behaviour, favouring organic products (Ruiz de Maya et al., 2011). Today there are hundreds of organic third-party certification programs and standards throughout the world. Organic certification is of major interest to NWFPs, because most of the standards consider organic both as products collected from the wild and semi-domesticated NWFPs, and at the same time, organic certification is well recognized and appreciated by the end users.

For the International Federation of Organic Agriculture Movements (IFOAM), the international umbrella organization of the organic products, organically wild harvested products shall:

> i) only be derived from a sustainable growing environment. Products shall not be harvested at a rate that exceeds the sustainable yield of the ecosystem, or threatens the existence of plant, fungal or animal species, including those not directly exploited; ii) operators shall harvest products only from a clearly defined area where prohibited substances have not been applied; iii) the collection or harvest area shall be at

> an appropriate distance from conventional farming or other pollution sources in order to avoid contamination; iv) the operator who manages the harvesting or gathering of common resource products shall be familiar with the defined collecting or harvesting area, including the impacts of collectors not involved in the organic scheme; v) operators shall take measures to ensure that wild, sedentary aquatic species are collected only from areas where the water is not contaminated by substances prohibited in these standards.
>
> (IFOAM, 2014)

The EU organic framework considers wild collection a sufficient action for obtaining the organic certification as well. This occurs if:

> i) the plants have grown naturally in natural areas, forests and agricultural areas, ii) in those areas have not, for a period of at least three years before the collection, received treatment with products other than those authorised for use in organic production [. . .], and iii) the collection does not affect the stability of the natural habitat or the maintenance of the species in the collection area.
>
> (European Union, 2007)

Organic certification does not mention the term NWFP, and does not specifically focus on forest ecosystems, but rather focuses on the quality of the land in which the product is sourced. Ecological specifications are included. Box 10.3 summarizes the approach of NWFP organic certification.

**Box 10.3 Organic certification (most of the standards)**

- *Specific to NWFPs*: does not directly target NWFPs as a category, but rather the collection of wild products, considering it as organic under some specifications;
- *Ecological specifications*: includes only general ecological specifications (e.g., "sustainable harvest");
- *Scale*: depends on the standard, but in general global.

### *10.3.4 Environmental performance standards*

Environmental performance certification aims at lowering the environmental impact of products, from a life cycle perspective. It does not specifically target NWFPs, but it can still award NWFPs that respect environmental performance criteria. "Ecolabels" are a sub-group of environmental labels; they are third-party certified and respond to special criteria of comprehensiveness, independence and reliability (UNOPS, 2009). The European Union Ecolabel[3] is an example of a regional ecolabelling scheme, coming from public initiative. The EU Ecolabel has been applied to some NWFPs, namely cork and cork products, such as coverings and panels. Box 10.4 summarizes EU Ecolabels for NWFPs.

**Box 10.4 Environmental performance certification (the EU Ecolabel example)**

- *Specific to NWFPs*: does not directly target NWFPs, and it does not use terminology for NWFPs nor for the collection of wild products;
- *Ecological specifications*: not included;
- *Scale*: EU countries.

### 10.3.5 Quality and food safety standards

Quality control and food safety certification aim at assuring the proper preparation of the products that enter in the market. The International Standard Organization (ISO) develops the most important standards in this sector. Of particular importance is the group of ISO 9001[4] standards, which addresses aspects of quality management, and, specifically for food products, the group of ISO 22000[5] standards, which addresses food safety management along the entire supply chain. An example of a safety certification program based on the principles of the ISO 9001 standard, as well as on the requirements of the Codex Alimentarius system, is the BRC Global Standards.[6] Edible NWFPs can be awarded by this standard, like in the case of "wild rice", which is an annual aquatic seed of *Zizania aquatica* found mostly in the freshwater lakes of Canada. Box 10.5 summarizes the approach of ISO to NWFPs.

**Box 10.5 Quality and food safety certification (ISO example)**

- *Specific to NWFPs*: does not directly target NWFPs, and it does not use terminology for NWFPs nor for the collection of wild products;
- *Ecological specifications*: not included;
- *Scale*: global.

A special type of quality certification is based on the Good Agricultural and Collection Practices (GACP) guidelines, published by the World Health Organization (WHO). WHO developed these technical guidelines for sustainable harvesting of plants. This model can be adapted at the national and regional levels. Similarly to GACP, there is also certification based on Good Manufacturing Practices guidelines for facilities, personnel and processing procedures for herbal medicines and wildcrafter guidelines (Shanley et al., 2008). Box 10.6 summarizes the approach of WHO guidelines to NWFPs.

**Box 10.6 Good Agriculture and Collection Practices certification (WHO guidelines example)**

- *Specific to NWFPs*: does not target NWFP category, targets instead herbs and medicinal plants;
- *Ecological specifications*: includes only general ecological specifications;
- *Scale*: global, can be adapted at the national and regional levels.

### *10.3.6 Socio-economic certification standards*

Socio-economic certification refers to the schemes that have social and economic focus. One of these is based on the Fairtrade standards. It aims at ensuring fair prices and empowering producers in the poorest countries of the world. Standards include requirements for environmentally friendly agricultural practices, such as safe use of agrochemicals, waste management, maintenance of soil fertility and water resources (Fairtrade Labelling Organizations International, 2011).

Fairtrade does not specifically target NWFPs. However, several NWFPs and products containing NWFPs have been certified, such as herbs, herbal teas, spices, juices and honey. For each category of product, a specific standard was developed. However, these standards do not include detailed ecological specifications. Box 10.7 summarizes the approach of Fairtrade to NWFPs.

**Box 10.7 Socio-economic certification (the Fairtrade example)**

- *Specific to NWFPs*: does not directly target NWFPs as a category, but it has standards for many subcategories of products (such as herbs, herbal teas and spices);
- *Ecological specifications*: not included;
- *Scale*: global, but mostly refers to the poorest countries of the world; it requires individual approval of each product.

### *10.3.7 Origin, geographical indications and traditional specialities certification*

Some standards and certification schemes apply to products with a recognizable traditional identity. In the EU, an example of such certification schemes comes from the public initiative. According to the EU Regulation 509/2006, three EU schemes promote and protect the names of quality agricultural products and foods: Protected Designation of Origin (PDO), Protected Geographical Indication (PGI) and Traditional Speciality Guaranteed (TSG) (European Union, 2006). This type of certification does not specifically target NWFPs nor wild collection. However, there are several NWFPs labelled with this certification. Although the framework is common for all EU countries, some countries more than others use this type of scheme. For example, overall, Italy counts a large number of registered (or under registration) products, at 318. Many of these are NWFPs or products made with NWFPs, as illustrated in Table 10.2. Box 10.8 summarizes the approach of origin, geographical indications and traditional specialities certification specific to NWFPs.

**Box 10.8 Origin, geographical indication and traditional speciality (the EU framework example)**

- *Specific to NWFPs*: does not specifically target the NWFP category nor the collection of wild production, and it does not use terminology for NWFPs;
- *Ecological specifications*: not included;
- *Scale*: EU countries.

## 10.4 Conclusions

The promotion of NWFPs can pass through major business practices and market-based instruments such as standards and certification. Several certification standards may be applied to NWFPs, which cover, in different degrees, the spheres of ecological, economic and social sustainability. Certification is not only an instrument to differentiate against industrial mass products, it is also a tool for tracing products. This can help in providing more transparency along the supply chain, supporting harvesting and exploitation done in a legal manner. It has the potential to ensure healthy employment conditions of the people involved in all the processes while respecting traditional use rights of local populations.

Since the aim of certification is to link the producer and end user through commonly understood terms of reference, the absence of a unified definition

*Table 10.2* NWFP certified under the EU geographical indications and traditional specialities scheme in Italy

| *Name* | *Description* | *Category of product and N°* | *Product name* |
|---|---|---|---|
| **Protected Designation of Origin – PDO** | Covers agricultural products and foodstuffs that are produced, processed and prepared in a given geographical area using recognized know-how. | Honeys (3) | Miele Varesino, Miele delle Dolomiti Bellunesi, Miele della Lunigiana |
| | | Chestnuts and chestnut flour (5) | Marrone di Caprese Michelangelo, Marrone di San Zeno, Castagna di Vallerano, Farina di castagne della Lunigiana, Farina di Neccio della Garfagnana |
| | | Pistachios (1) | Pistacchio Verde di Bronte |
| | | Hazelnuts (1) | Nocciola Romana |
| **Protected Geographical Indication – PGI** | Covers agricultural products and foodstuffs closely linked to the geographical area. At least one of the stages of production, processing or preparation takes place in the area. | Chestnuts (11) | Marroni del Monfenera, Marrone del Mugello, Marrone di Serino, Marrone di Combai, Marrone della Valle di Susa, Marrone di Roccadaspide, Castagna Cuneo, Castagna del Monte Amiata, Marrone del Mugello, Castagna di Montella, Marrone di Castel del Rio |
| | | Hazelnuts (2) | Nocciola del Piemonte Nocciola di Giffoni |
| | | Mushroom (1) | Fungo di Borgotaro |
| **Traditional Speciality Guaranteed – TSG** | Highlights traditional character, either in the composition or means of production. | – | – |

Source: Own elaboration

may have consequences for its success. We showed that the term "NWFP" is only adopted in forest standards, while other standards, such as the organic standards, prefer wording containing the term "wild". The use of the latter seems to be a more efficient marketing strategy compared to the NWFP acronym, because it is more easily understood by the end user.

Among the certification types of major interest for NWFPs, only three explicitly target NWFPs or "gathered from the wild". Schemes such as Sustainable Forest Management Certification, wild certification and organic certification look at the harvesting stage of supply chains and also include ecological specifications for sustainable harvesting (detailed specifications in the cases of Sustainable Forest Management Certification and wild certification and only general specifications in the case of organic certification). Only in these cases, the economic actors give signals that the ecological impact of the NWFPs harvesting is positive, or at least not negative.

## Notes

1 This chapter has been in part reproduced from the following works of the authors:

Corradini, G. and Pettenella, D., (2017). *Promoting NWFPs: Branding, Standards and Certification in COST Book of Non-Wood Forest Products*, in preparation.

Corradini, G. and Pettenella, D., (2017). WFP in Europe: Marketing tools. In Prokofieva, Mavsar, Wolfslehner (eds.) *What Science Can Tell Us on Non-Wood Forest Products*. European Forest Institute, Joensuu, Finland.

2 FairWild Fundation. URL: www.fairwild.org/

3 Introduced by the Regulation (EC) No 880/92 and amended by Regulation (EC) No 1980/2000 and Regulation (EC) No 66/2010, http://eur-lex.europa.eu/legal-content/EN/TXT/?uri=URISERV:co0012

4 ISO 9001 Quality Management standard. URL: https://www.iso.org/iso-9001-quality-management.html

5 ISO 22000 Food Safety Management standard. URL: https://www.iso.org/iso-22000-food-safety-management.html

6 BRC Global Standards. URL: https://www.brcglobalstandards.com

## References

AS (2007). *The Australian Forestry Standard*. Traffic International, Cambridge, UK.

BioLand (2013). *Bioland Standards*. Traffic International, Cambridge, UK.

Bio Suisse (2014). *Standards for the Production, Processing and Marketing of Bud Products*. Traffic International, Cambridge, UK, pp. 247–270.

Burgener, M., Walter, S. (2007). Trade Measures-Tools to Promote the Sustainable Use of NWFP? An Assessment of Trade Related Instruments Influencing the International Trade in Non-Wood Forest Products and Associated Management and Livelihood Strategies. Non-Wood Forest Products Working Document (6), FAO, Rome, Italy.

CFCC (2014). *Forest Certification in China-Non Timber Forest Product Certification Audit Directive*. China Forest Certification Council. Beijing, China.

CSA (2013). *Sustainable Forest Management*. Traffic International, Cambridge, UK.

European Union (2006). Council Regulation (EC) No 509/2006 of 20 March 2006 on Agricultural Products and Foodstuffs as Traditional Specialities Guaranteed.

European Union (2007). Council Regulation (EC) No. 834/2007 of 28 June 2007 on Organic Production and Labelling of Organic Products and Repealing Regulation (EEC) No. 2092/91.

Fairtrade Labelling Organization International (2011). *Fairtrade Standard for Herbs, Herbal Teas & Spices for Small Producer Organizations*. Fairtrade Labelling Organization International, Bonn, Germany.

FairWild Foundation (2010). *FairWild Standard Version 2.0*. Weinfelden, Switzerland, Weinfelden, Switzerland.

FairWild Foundation (2017). *FairWild Certified Ingredients under Production*. Fair Wild Foundation, Switzerland.

FAO (1999). Non-Wood Forest Products and Income Generation. *Unasylva* 198, 1–77.

FSC (2014). *FSC Principles and Criteria for Forest Stewardship*. Traffic International, Cambridge, UK.

FSC (2017). *The FSC National Forest Stewardship Standard of Italy (FSC-STD-ITA-01–2017 V 1–0)*. FSC Global Development GmbH, Bonn, Germany.

IFOAM (2008). Definition of Organic Agriculture in English, as Approved by the IFOAM General Assembly in Vignola, Italy in June 2008.

IFOAM (2012). *The IFOAM Norms for Organic Production and Processing*. Traffic International, Cambridge, UK.

IFOAM (2014). The IFOAM Norms for Organic Production and Processing Version 2014. International Foundation for Organic Agriculture, Bonn, Germany.

IMO/SIPPO (2005). *Guidance Manual for Organic Collection of Wild Plants*. Institute for Marketecology. Weinfelden, Switzerland

ITC International Trade Center (2007). *Overview of World Production and Marketing of Organic Wild Collected Products*. International Trade Centre UNCTAD/WTO, Geneva, Switzerland.

Kilchling, P., Hansmann, R., Seeland, K. (2009). Demand for Non-Timber Forest Products: Surveys of Urban Consumers. *Forest Policy and Economics*, 11(4), July 2009, 294–300, ISSN 1389–9341, http://dx.doi.org/10.1016/j.forpol.2009.05.003.

Lange, D. (1998). *Europe's Medicinal and Aromatic Plants: Their Use, Trade and Conservation*. Traffic International, Cambridge, UK.

MAFF (2012). *Japanese Agricultural Standards (JAS) for Organic Plants and Organic Processed Foods*. Traffic International, Cambridge, UK.

Naturland (2014). *Naturland Standards on Production*. Traffic International, Cambridge, UK.

NEPCon (2014). *FSC Global Non Timber Forest Product Certification Addendum*. NEPCon, Copenhagen, Denmark.

NEPCon (2017). *FSC Non Timber Forest Product Certification Addendum for Russia*. NEPCon, Copenhagen, Denmark.

Pierce, A.R., Shanley, P., Laird, S.A. (2003). Certification of Non-Timber Forest Products: Limitations and Implications of a Market-Based Conservation Tool. Paper presented at the International Conference on Rural Livelihoods, Forests and Biodiversity 19–23, May 2003. Bonn, Germany.

Pierce, A.R., Shanley, P., Laird, S.A. (2008). Non Timber Forest Products and Certification: Strange Bedfellows. *Forests, Trees and Livelihoods*, 18(1), 23–35.

Ruiz de Maya, S., López-López, I., Munuera, J.L. (2011). Organic Food Consumption in Europe: International Segmentation Based on Value System Differences. *Ecol. Econ*. 70, 1767–1775. doi:10.1016/j.ecolecon.2011.04.019

SA (2014). *Soil Association Organic Standards, Farming and Growing*. Traffic International, Cambridge, UK.

SFI (2015). *SFI 2015–2019: Standards and Rules.* Traffic International, Cambridge, UK.
Shackleton, S., Shackleton, C., Shanley, P. (2011). *Non-Timber Forest Products in the Global Context.* Traffic International, Cambridge, UK.
Shanley, P., Pierce, A.R., Guillén, A., Laird, S.A. (2002). *Tapping the Green Market: Certification and Management of Non-Timber Forest Products.* Earthscan, London, UK.
Shanley, P., Pierce, A.R., Laird, S.A., Robinson, D. (2008). *Beyond Timber: Certification and Management of Non-Timber Forest Products.* CIFOR (Center for International Forestry Research), Bogor, Indonesia.
UEBT (2012). *Union for Ethical BioTrade.* Traffic International, Cambridge, UK.
UNOPS (2009). *A Guide to Environmental Labels: For Procurement Practitioners of the United Nations System.* UNOPS.
USDA (2011). *Guidance Wild Crop Harvesting.* Traffic International, Cambridge, UK.
Vantomme, P., Walter, S. (2003). Opportunities and Challenges of Non-Wood Forest Products Certification. Paper submitted to the World Forestry Congress, Quebec, 21–28 September, Rome, Italy.
Walter, S. (2002). Certification and Benefit-Sharing Mechanisms in the Field of Non-Wood Forest Products: An Overview. *Med. Plant Conserv. Newsl. IUCN Species Surviv. Comm. Med. Plant Spec. Gr.* 8.
Walter, S. (2006). Certification of Non-Wood Forest Products: Relevant Standards, Preliminary Experiences and Lessons Learnt. Paper presented at the 1st International Conference on Wild Organic Production Teslic, Bosnia Herzegovina, 3–4 May 2006, 3–4.
Willer, H., Lernoud, J. (Eds.) (2017). The World of Organic Agriculture Statistics and Emerging Trends 2017. Research Institute of Organic Agriculture (FiBL), Frick, and IFOAM – Organics International, Bonn. Version 1.3 – February 20, 2017.

Chapter 11

# Fair Trade certification of forest products

*Alessandro Leonardi*

This chapter provides an overview of Fair Trade certification initiatives for forest products, focusing in particular on the pilot projects and applications integrating the Forest Stewardship Council (FSC) certification with Fair Trade. The first paragraph (paragraph 11.1) provides a background on existing Fair Trade standards and introduces the main links and pilot initiatives between forest certification and Fair Trade schemes, namely the FSC–Fairtrade (one of the leading international standards for Fair Trade) dual certification pilot project (paragraph 11.2) and the FSC Small and Community Label Option (paragraph 11.3).

## 11.1 Fair Trade system and standards applicable to forest products

Since colonial times, large-scale industrial concessions and plantations have been the main forestry models. These models have neither benefitted poor and fragile communities nor achieved environmental conservation and local development goals (Angelsen and Wunder, 2003). On the contrary, sometimes industrial forest management models led to illegal logging and widespread corruption. In any case, almost one-quarter of the global forest area is owned and/or managed by communities, especially in tropical countries, while decentralized and community-based forest governance is rapidly expanding (Agrawal et al., 2008). Governments, non-governmental organizations (NGOs) and international organizations are promoting community-based forest development strategies in order to provide "better forestry, less poverty" (FAO, 2006). However, despite their growing role, few examples of sustainable and economically viable Forest Enterprises involving communities exist. This is partly due to complexities of the socio-political and environmental contexts in which communities live, and the difficulties involved in linking community products with markets (Humphries and Kainer, 2006).

Forest certification is still considered one of the most recognized market-based instruments to promote sound forest management worldwide. Certification in forestry has also become a sort of development approach for communities to make claims for the social and environmental values of their enterprises (Donovan et al., 2006). Community forestry is seen as more economically viable, more resilient and with stronger social and environmental positive impacts than large private enterprises (Antinori and Barton, 2005). However, in order to be more competitive and to distinguish products within the market, it is reasonable for communities to make claims for the social and environmental value of their activities through forest certification and labelling of forest products (Macqueen, 2009). Nevertheless, if we consider the geographical distribution of certified forests and forest certification schemes, despite the great diffusion in the Northern Hemisphere, most of the tropics are lacking any kind of national or international third-party forest certification initiative (UNECE/FAO, 2015). Moreover, forest certification has been relegated to large-scale industries, mainly in North America and Europe, while social and environmental issues concerning the world's forests arise mainly in the tropics, especially with indigenous people and worker's rights (Tikina, 2008). Therefore, if we consider community forests (CFs) and Small and Medium Forest Enterprises (SMFEs) and their associations, the importance of ownership in the tropical countries and the role of these businesses in reducing forest-related poverty, increasing the access to certification for them should be a priority in order to gain more responsible and viable certified forests in the tropics. Despite some efforts, all around the world CFs and SMFEs are facing many barriers to becoming certified due to the high cost and complexity of certification, lack of market channels, lack of premium prices and supply scale and quality issues (Higman and Nussbaum, 2002; Taylor, 2005).

FSC® has tried to adapt its certification system to facilitate access to certification for CFs and SMFEs. So far, FSC seems to be the only forest certification standard-setting organization going in the direction of increasing access of small-scale forestry in the tropics, through the promotion of group certification, Small and Low Intensive Managed Forests (SLIMF), and other specific projects that are in the core of the FSC global strategy and social policy program (FSC, 2010). However, despite the great achievement in terms of simplified procedures and reduced costs for certification, market benefits were, and are, still very limited for community forests and small to medium sized enterprises.

In 2008, several international stakeholders supporting forest certification gathered together in workshops and asked for studies to be carried out to understand how forest certification schemes can distinguish CF and SMFE products in the market in order to increase market benefits. Attempting to develop additional forest certification schemes or standards when, internationally speaking, there are a multitude of them, seemed not to be recommendable. Therefore, linking the Fair Trade approach with forest certification seemed to be the most viable solution due to the international recognition of Fair Trade labels and concepts (Box 11.1), as well as for the premium price mechanism that could increase market benefits for smallholders. Two institutional initiatives resulted from this international discussion: the FSC–Fairtrade Dual Labelling Pilot Project and the FSC Small and Community Label Option.

### Box 11.1 Fair Trade systems

At the global level, the World Fair Trade Organization (WFTO) and the Fairtrade Labelling Organization (FLO) schemes try to respond to the needs of organizations of small producers, mainly with reference to the agriculture sector. WFTO is a membership-based organization, with a guarantee system that does not follow the ISO guidelines for product certification. The approach is more system-based and it has been structured in order to lower the cost of auditing and allow step-by-step access for marginalized producers. On the other hand, FLO is based on third-party standards and certification procedures that follow the ISO guidelines for product certification. Premium prices, long-term contracts, attention to workers' rights and gender equity are the core principles governing Fair Trade. Consumer demand for Fair Trade products is growing and it is stepping out of its niche and entering the mainstream market. In 2016, FLO counted 1,225 producer organizations in 74 countries and the sales of FLO certified products reached a record height of €7.88 billion (Fairtrade International, 2017). Some FTOs sell food products branded and distributed on a mass-market scale through do-it-yourself channels; however, this has been replicated in the wood-crafting sector in just a few cases. Timber could also be a Fair Trade product, but often no assurance of its sustainability is given by the Fair Trade value chain.

At the same time, some Fair Trade Organizations (FTO) are already linking and commercializing Fairtrade and FSC-certified products under the World Fair Trade Organization (WFTO) network. For example, since 2005, COPADE *Comercio Para el Desarrollo*, an FTO based in Spain, delivers Dual Certified Forest Products to the Spanish market to be sold through Fair Trade shops and private companies. The dual-certified FSC–Fairtrade products are derived from FSC-certified community forests and groups of small timber processors (UNICAF and COATLAHL) based in Honduras and Bolivia (Box 11.2). COPADE activities are part of sensibilization and promotional campaign called *Madera Justa*, i.e., Fair Timber. The campaign turned into a certification initiative, creating a platform in Spain and drafting a standard and certification system for fair and sustainable timber trade. Another example is FORCERT (Box 11.3), an FSC-certified community forest in Papua New Guinea that is also a Fairtrade organization certified by the WFTO. FORCERT is marketing its timber-based products to the Australian and European market. Another importing FTO that is FSC-certified and based in the UK is selling rubber-based dual-certified products (like condoms, footballs, gloves, balloons, etc.) coming from Pakistan and India.

**Box 11.2 COATLAHL: dual certification in Honduras**

The Regional Agroforestry Cooperative "Colón, Atlántida, Honduras" Ltda (COATLAHL) first obtained FSC certification in 1996. The cooperative holds a group FSC Certificate on behalf of 14 small community-based timber-producing groups, who manage 19,500 ha of natural broadleaf forests. COATLAHL provides marketing and sales services and has a furniture workshop with a certified supply chain. Their main products are sawn timber and solid wood furniture. The community benefits from group certification and also follows the streamlined procedures for Small and Low Intensive Managed Forests (SLIMFs); following these procedures provides a smooth way to reach and maintain the certification. The reason for this is that there are fewer criteria to be met for forest operations that are considered small or with low intensive extraction volume according to FSC standards. To increase benefits for forest community families and forest workers, COATLAHL, under a pilot project between FSC and FLO (Fairtrade Labelling Organization), decided to get Fairtrade certification. Gaining more market access and higher timber prices and responding to market demand were the main reasons to apply for Fairtrade certification. In the case of COATLAHL, which was already FSC certified, the Fairtrade certification was a tool that helped to strengthen its commitment toward socially and environmentally responsible forest management. Thanks to the technical support of NGOs, such as Bosque del Mundo, and Fair Trade importers, part of the timber processed by COATLAHL is being sold in Spain and Denmark.

For more information: www.coatlahl.com

**Box 11.3 FORCERT: dual certification in Papua New Guinea**

FORCERT is a not-for-profit company based in Papua New Guinea that promotes sustainable forest management. FORCERT provides certification and marketing services under both FSC and Fair Trade to local small-scale producers. FORCERT uses Forest Stewardship Council (FSC) group certification as a management, marketing and networking tool. It links community Forest Enterprises to central timber yards, and combines the output of these yards to reach overseas markets. FORCERT adopts a stepwise approach to certification. To join the group, the first requirements are the respect of legality and Fair Trade principles. Then, within a scheduled time plan, the new communities are pushed to move toward Controlled Wood and, finally, to fulfill FSC Forest Management Certification. FORCERT believes in a fair and transparent independently certified trade of forest products that recognizes the important role of local landowners and ensures the different values of their forests are appreciated and maintained. Thus, FORCERT is an FSC-certified forest operation and a Fair Trade Organization verified by WFTO. This means that the timber pricing and sales arrangements of FORCERT have been recognized as meeting Fair Trade standards.

For more information: www.forcert.org.pg

## 11.2 FSC–FLO dual certification pilot project

In 2009, after some explorative studies, FLO and FSC launched a joint pilot project to help community-based and small-scale timber producers to gain market access while ensuring a fair price for their products. The project aims to test the concept of dual labelling (FSC and FLO) for smallholders and communities in developing countries. Within the pilot, FLO has developed a new Fairtrade standard for timber for Forest Enterprises that source from small-scale/community-based producers, which is designed to complement the FSC Principles and Criteria for Forest Management for small and/or community-based forestry operations. The Fairtrade standard for timber is targeted to:

- Forest Enterprises that already hold a valid FSC combined Forest Management/Chain of Custody (FM/CoC) certificate that are themselves, or source their timber from, small-scale or community-based producers (for the definition see Box 11.4);
- traders in the supply chain that subsequently handle the Fairtrade products and who must also already have a valid FSC Chain of Custody (CoC).

The Fairtrade payer is the buyer responsible for paying the Fairtrade Minimum Price and the Fairtrade Premium to the Forest Enterprise.

**Box 11.4 Definition of small-scale or community-based producers from "labelling products from small and community producers" (FSC-ADV-50-003 V1-0 EN)**

**Definition of Small Producer:** a Forest Management Unit (FMU) or group of FMUs that meet(s) the SLIMF eligibility criteria (FSC-STD-1-003a) and Addenda. For group certificate holders that include non-SLIMF FMUs, only the FMUs categorized as SLIMFs are eligible.

**Definition of Community Producer:** A Forest Management Unit complying with the following tenure AND management criteria:

- **Tenure:** the legal right to manage a Forest Management Unit (e.g., title, long-term lease, concession, etc.) is held at the communal level, AND the community members must be either indigenous peoples or traditional people, OR the Forest Management Unit meets the SLIMF eligibility criteria;
- **Management:** The community actively manages the Forest Management Unit through a concerted effort (e.g., under a communal forest management plan) OR the community authorizes management of the forest by others (e.g., resource manager, contractors, forest products company).

If the community authorizes management of the forest by others, criterion 1 and either criterion 2 or 3 must be met:

1 The community's own representative institution has legal responsibility for the harvesting operations, AND
2 The community performs the harvesting operations, OR
3 The community's own representative institution is responsible for the forest management decisions and follows and monitors the operations.

The Fairtrade standard for timber is divided into four main sections: social, economic, environmental development and labour conditions. Each section is then sub-divided into principles and several requirements. The latter are classified into:

- general, which all Forest Enterprises must meet from the moment they join Fairtrade;

- minimum, which must be met before the first audit; and
- progress requirements, against which Forest Enterprises must demonstrate compliance over time and by means of continuous improvement.

The standard introduces the concept of Fairtrade Association (which is not necessarily but can be identical with the Forest Enterprise), which represents small forest owners or community forests, subcontractors and forest workers. The association is in charge of managing the premium price in order to allow benefit-sharing among all stakeholders. The premium is managed directly by the Fairtrade Committee, which is elected by the members of the Fairtrade Association. Besides the social, economic, environmental and labour conditions, the standard defines clear trading rules for contracts among the supply chain actors, traceability, sourcing plans, pre-financing, pricing, labelling and packaging. The traceability section defines additional CoC requirements that apply to all operators down the supply chain, establishing the physical separation of Fairtrade and non-Fairtrade timber and describing a percentage system by volume labelled as Fairtrade.

The sourcing plans describe the qualities, quantities, dates of delivery or purchase and the price or value of Fairtrade-Certified Wood products that the buyer expects to purchase throughout the year. This is particularly important for small Forest Enterprises, which are often unable to face unexpected purchasing orders. It also helps to match sustainable forest management activities with buyers' requirements by, for example, planning felling activities in advance by type of species and related volumes.

The core Fairtrade principles are related to:

- Pre-financing; traders are encouraged to offer other financial assistance to the Forest Enterprise or other forms of financing such as "prepayment" and/or "advance-payment" to producers. This is particularly important to face the sustainable forest management costs and felling activities and long-term investments that are required in forestry.
- Minimum price; the price paid for Fairtrade timber products must cover the Costs of Sustainable Production, which are the costs for producing wood products according to sustainable and fair criteria, calculated by the producers themselves.
- Premium price; the Forest Enterprise, the Fairtrade Committee and Fairtrade payers should agree a premium level of 5% of the value of estimated annual sales of Fairtrade. The premium adds on the minimum price and is based on a negotiation between the actors involved. The payment of the premium can be delivered in two models:
    - direct payment model, directly from the Fairtrade Licensee to the Forest Enterprise or via a trusted intermediary agreed by both parties;

- pass through model, in which the Fairtrade Premium is specifically itemized on the invoices and passed through the trade chain from the Forest Enterprise to the Fairtrade Licensee without elaboration.

The final result that the consumer experiences with their product is a dual label, FSC + FLO, with a payoff that says: "FSC and Fairtrade together for forests and communities". The marketing of these community forest products is also coupled with another tool that FSC has developed alongside the dual labelling FSC/FLO pilot: the FSC Small and Community Label Option (see paragraph 11.1.2).

The standard has been tested on community forestry operations in Honduras (COATLHAL) and a group of small/low intensity producers in Chile. Another group of small/low intensity producers was assessed for Fairtrade timber certification in Bolivia, however, it never obtained the full certification. All of these producers were previously FSC certified. To date, only one FLO/FSC supply chain is still working, that is, the Chilean pilot (SSC Forestry) and Khars Sweden, providing hardwood flooring for the Swedish market. All other pilots have failed due to the cost of certification and technical and market support, which was initially covered by development NGOs, the lack of assistance for actors along the value chain to manage the sourcing from smallholders and the lack of buyers willing to pay a premium price for dual labelled FLO/FSC timber (FSC, 2015). The initial project ended in 2013 but, according to FSC and Fairtrade, a second phase is expected to be launched.

## 11.3 FSC Small and Community Label

FSC has developed an Advice Note for Labelling Products from Small or Community Producers (SCLO) to introduce a special label to differentiate products originating from FSC-certified community forests and SLIMF producers around the globe. The main issue raised during the development of the document and the contribution to the FSC network is the new definition of "community producers",[1] which was drafted through a large participatory process. According to the advice note, a community producer is a Forest Management Unit complying with both the following criteria:

- Tenure: the legal right to manage a Forest Management Unit (e.g., title, long-term lease, concession) is held at the communal level, AND
  - the community members must be either indigenous peoples or traditional peoples, OR
  - the Forest Management Unit meets the SLIMF eligibility criteria.

- Management: The community actively manages the Forest Management Unit through:
  - a concerted effort (e.g., under a communal forest management plan), OR
  - a management authorization by the community to others (e.g., resource manager, forest products company).

The initiatives enable community forests and SLIMF producers to use a special claim on their FSC on-product label: "From well-managed forests of small or community producers". The advice note then introduces a new FSC product group called FSC SCLO that will have to be physically separated by other FSC or non-FSC materials. The claim can be used with both FSC 100% and Mix labels. However, if using FSC Mix, the Controlled Wood in the mix must also be from certified small and community sources. Therefore, in both cases the products have to be 100% SCLO sourced.

An SCLO Marketing Toolkit supports retailers to launch campaigns (following the Fair Trade approach) that promote these products by telling consumers the story of producers. The campaign is called the "Made with Heart" story. This creates a connection between the consumers and the producers: by purchasing these products, consumers help to improve the livelihood of these forest communities. An SCLO web marketplace also supports producers worldwide to market their products and status as SCLOs. So far, few producers have adopted the option, nevertheless the label has a great potential for its high-level marketing tools and campaigns that create market access at no additional costs for forest managers.

## Note

1 Note that the SLIMF definition and eligibility criteria were already defined in FSC-STD-1-003a.

## References

Agrawal, A., Chatre, A., Hardin, R. (2008). Changing Governance of the World's Forests. *Science (80) 320*, 1460.

Angelsen, A., Wunder, S. (2003). Exploring the Forest-Poverty Link: Key Concepts, Issues and Research Implications., Occasional Paper: CIFOR. Center for International Forestry Research., Jakarta, Indonesia.

Antinori, C., Barton, D. (2005). Community Forest Enterprises as Entrepreneurial Firms: Economic and Institutional Perspectives from Mexico. *Livelihoods, For. Conserv. 33*, 1529–1543.

Donovan, J., Stoian, D., Macqueen, D., Grouwels, S. (2006). The Business Side of Sustainable Forest Management: Small and Medium Forest Enterprise Development for Poverty Reduction. *Nat. Resour. Perspect. 104*.

Fairtrade International (2017). Creating Innovations, Scaling Up Impact: Annual Report 2016–2017. Available at https://annualreport16-17.fairtrade.net/en/

FAO (2006). Better Forestry, Less Poverty: A Practitioner's Guide. FAO For. Pap. (149), Rome.

FSC (2010). FSC Certification and Social Issues. FSC Fact Sheet. Forest Stewardship Council, Bonn, Germany.

FSC (2015). The FSC: Fairtrade Dual Certification Pilot Project. Forest Stewardship Council, Bonn, Germany.

Higman, S., Nussbaum, R. (2002). *How Standards Constrain the Certification of Small Forest Enterprises*. ProForests, Oxford

Humphries, S., Kainer, A. (2006). Local Perceptions of Forest Certification for Community-Based Enterprises. *For. Ecol. Manage. 235*, 30–43.

Macqueen, D. (2009). Roots of Success: Cultivating Viable Community Forestry. IIED Briefing Paper, London.

Taylor, P.L. (2005). A Fair Trade Approach to Community Forest Certification: Framework for Discussion. *Certifying Rural Spaces Qual. Prod. Rural Gov. 21*, 433–447.

Tikina, A.V. (2008). Forest Certification: Are We There Yet? *Perspect. Agric. Vet. Sci. Nutr. Nat. Resour. 3*, 8.

UNECE/FAO (2015). *Forest Products Annual Market Review 2014–2015*. United Nations, Geneva.

Chapter 12

# At the frontier

## Towards ecosystem services certification

*Lucio Brotto, Alex Pra, Ariadna Chavarria and Alessandro Leonardi*

This chapter introduces the emerging initiatives for the certification of ecosystem services, focusing on the FSC certification of ecosystem services (paragraph 12.2) and other certification initiatives integrating ecosystem services other than carbon, such as biodiversity and water-related services (paragraph 12.3).

### 12.1 Certifying ecosystem services

Ecosystem services are defined as "the benefits people receive from the planet's ecosystems" (MEA, 2005, p. 3). Forests play an important role in providing essential ecosystem services for human well-being, such as biodiversity conservation, carbon sequestration and storage, regulating water cycle and quality, soil conservation, recreational services and so on (Table 12.1). The increasing demand for food, raw material and energy, resulting in unsustainable forest management practices, conversion of forests to agriculture and forest degradation processes, are threatening ecosystems and their capacity to provide ecosystem services. The Millennium Ecosystem Assessment estimated that about 60% of ecosystem services are degraded or heavily compromised (MEA, 2005).

In recent years, several mechanisms have been tested and put in practice to give value to ecosystem services in order to improve their management and protect ecosystems, from more traditional ones such as regulation of land use, taxes and subsidies and programs to support community forestry, to more innovative ones such as Payments for Ecosystem Services (PES) and the use of voluntary certification schemes. PES mechanisms have gained prominence in recent years, emerging as a promising instrument. In practice, a PES mechanism consists of paying, or alternatively providing non-monetary rewards, to forest owners or communities to manage the ecosystem on their land in a way that maintains or enhances ecosystem service provision (a detailed definition is provided in Box 12.1).

*Table 12.1* Forest ecosystem services

| *Category* | *Forest-based ecosystem services* |
|---|---|
| Cultural services | • Aesthetics and landscape beauty<br>• Cultural values and symbolism<br>• Educational opportunities<br>• Recreational activities<br>• Spiritual enrichment<br>• Tourism |
| Provisioning services | • Fish (from coastal forests or mangroves)<br>• Traditional medicines<br>• Food, fuelwood and timber<br>• Water supply |
| Regulating services | • Carbon sequestration<br>• Climate regulation and stabilization<br>• Control of pests that affect plants or animals<br>• Decomposition of waste<br>• Disease control<br>• Erosion control<br>• Improvements in air quality<br>• Maintenance of regional precipitation patterns<br>• Mitigation of floods and droughts<br>• Moderation of the force of winds and waves<br>• Protection from the sun harmful ultraviolet rays<br>• Water purification |
| Supporting services | • Biodiversity conservation<br>• Dispersal of seeds<br>• Maintenance and renewal of soils and soil fertility<br>• Maintenance of habitats for plants and animals<br>• Pollination of crops and natural vegetation<br>• Translocation of nutrients |

Source: Adapted from Millennium Ecosystem Assessment (2005)

**Box 12.1 Definition of Payment for Ecosystem Services (PES)**

A large body of literature describes PES as

> (a) a voluntary transaction where (b) a well-defined environmental service (ES) or a land use likely to secure that service (c) is being 'bought' by a (minimum one) service buyer (d) from a (minimum one) service provider (e) if and only if the service provider secures service provision (conditionality).
>
> (Wunder, 2005)

In more detail, the five specific elements that compose a PES are:

- The voluntariness of the transaction: the degree to which the contracting parties, the service provider(s) and the beneficiary(ies) enter in agreement and participate through a free and informed process of negotiation. The voluntariness principle is therefore a characteristic that differentiates PES from the more "government-based" command and control measure.
- Well-defined ecosystem services: the transaction and thus the whole PES design should be based on a "well-defined" ecosystem service, which would be the subject of the contract. Specific metrics and monitoring processes and output indicators shall be identified in order to verify the type of land use likely to secure the service and to measure the final service provided to beneficiaries.
- Actors involved: although the PES theory mainly refers to two actors (a service provider and a service beneficiary), other actors can influence the design and implementation of the contractual agreement. We can therefore summarize the main groups that are typically involved in a PES scheme:
  - Buyers or beneficiaries: those who are willing to pay for an improved, safeguarded or restored ecosystem service. These include citizens, water utilities, municipalities, beverage companies, etc.
  - Sellers or service providers: land and resource managers whose change of management practice can potentially secure or improve supply of the ecosystem service.
  - Intermediaries: those who can serve as agents linking buyers and sellers and can help with scheme design and implementation. They are often NGOs, parks authorities, river trusts, farmers associations, etc.
  - Knowledge providers: these include resource management experts, valuation specialists, land use planners, universities, participation experts and business and legal advisors who can provide knowledge essential to scheme development.
  - Regulators: those who can impose command and control measures that influence PES or can regulate and/or facilitate the start-up and the effectiveness of PES mechanisms.
- Conditionality of the payment: the degree to which the service provision is conditional to the payment. This principle is often very hard to meet because of several factors. In many cases there is a lack of knowledge about the "baseline" scenario, so as to understand and measure how the payment has influenced the service provision, compared to a "no intervention" scenario.

(Wunder, 2005)

Despite these well known definitions of PES, PES as an acronym is still used in the literature without a clear and standardized definition, leading to many interpretations and conceptualizations.

Recently, forest certification schemes and other non-governmental organizations have supported the idea of developing new systems for certifying the contribution of forests in relation to the provision of ecosystem services. Certification of ecosystem services can be defined as a market-based mechanism that includes activities meant to guarantee that a given forest stand is explicitly managed in a way that maintains or enhances the provision of ecosystem services. According to Jaung and Putzel (2013), certifying ecosystem services can help overcome many of the challenges facing PES markets, for example, it can:

- reduce transaction costs for buyers and sellers;
- improve communication and increase transparency between buyers and sellers (e.g., thanks to third-party audits, certified claims, etc.);
- improve the ability of forest owners or managers to measure and record key indicators of ecosystem service protection and provision, as well as strengthen the monitoring, reporting and verification of the outcomes of the activities;
- safeguard the interests of local communities and indigenous people.

Some ecosystem services have already established certification and verification schemes, i.e., carbon, where measuring scientific tools are more mature and relatevely more simple than other ecosystem services (see Chapter 8). However, there are few certification standards and systems in place for certifying forest ecosystem services other than carbon. There are some approaches to directly or indirectly incorporating specific ecosystem services in certification schemes for forest management and for carbon, i.e., the inclusion of biodiversity protection as a co-benefit in the Gold Standard and in the Climate, Community and Biodiversity (CCB) Standard, or the inclusion inside corporate responsibilities strategies such as the Business and Biodiversity Offset Partnership (BBOP) initiative for biodiversity and the Water Restoration Certification initiatives for water restoration. These initiatives concentrate on specific ecosystem services; the only initiative to date trying to expand the approach to all ecosystem services is led by FSC®. Through its Forest Certification for Ecosystem Services (ForCES) project, new indicators to incorporate forest ecosystem services in Forest Management Certification are being developed.

## 12.2 FSC Forest Certification for Ecosystem Services

Protecting ecosystem services has always been part of FSC Forest Management Certification. However, in recent years FSC recognized the need for futher incentives to manage forests responsibly, valuing not only forest products but also forest ecosystem services. With the aim of adapting FSC standards to ecosystem service markets, FSC launched the project Forest Certification for Ecosystem Services (ForCES)[1] in 2011. The ForCES project is aimed at designing and testing an adapatation of the FSC certification system that could lead to new rewards for forest owners and managers from emerging markets for ecosystem services. The project was supported by the Global Environmental Facility (GEF) and the United National Environment Program (UNEP). Ten pilot tests were lauched in Chile, Indonesia, Nepal and Vietnam, testing five different forest ecosystem services. The aim of the pilot tests was to assess the actual market demand for ecosystem services and develop a new set of national standard requirements to include within already existing FSC Forest Management Certification procedures. The new sets of indicators were developed to verify the impacts of five ecosystem services: carbon sequestration and storage, biodiversity conservation, watershed services, soil conservation and recreational services.

The strategic role that ecosystem services will play in the future of FSC has been highlighted with the updates of FSC's policy framework and strategies (e.g., Global Strategic Plan 2015–2020 and the FSC Ecosystem Services Strategy) and revising its normative framework (standards, policies and procedures). The FSC Principles and Criteria (FSC-STD-01–001 V4) have been revised to include, among other changes, explicit references to ecosystem services, i.e., Principle 6 requires the organization to "maintain, conserve and/or restore ecosystem services", and Criterion 5.1 requires the Forest Management Organization to "identify, produce, or enable the production of, diversified benefits and/or products, based on the range of resources and ecosystem services existing in the Management Unit". The main document that defines the incorporation of ecosystem services in Forest Management Certification is the new Annex C of the FSC International Generic Indicators (FSC-STD-60–004 V1–0), which were developed based on Version 5 of the FSC Principles and Criteria. This Annex lists all the additional requirements related to the maintainance and/or enhancement of ecosystem services (Figure 12.1).

To further streamline the process of benefiting from ecosystem services, in 2017, FSC developed the Ecosystem Services Procedure: Impact Demonstration and Market Tools FSC-PRO-30–006 V1–0 EN. The scope of the Ecosystem Services Procedure is to:

- allow FSC Forest Management Certificate holders to demonstrate the positive impact that their forest management has on ecosystem services;
- allow FSC Chain of Custody certificates to be used for FSC's ecosystem services claims;
- clarify how certification bodies shall evaluate FSC Certificate holders' compliance with the Ecosystem Services Declaration.

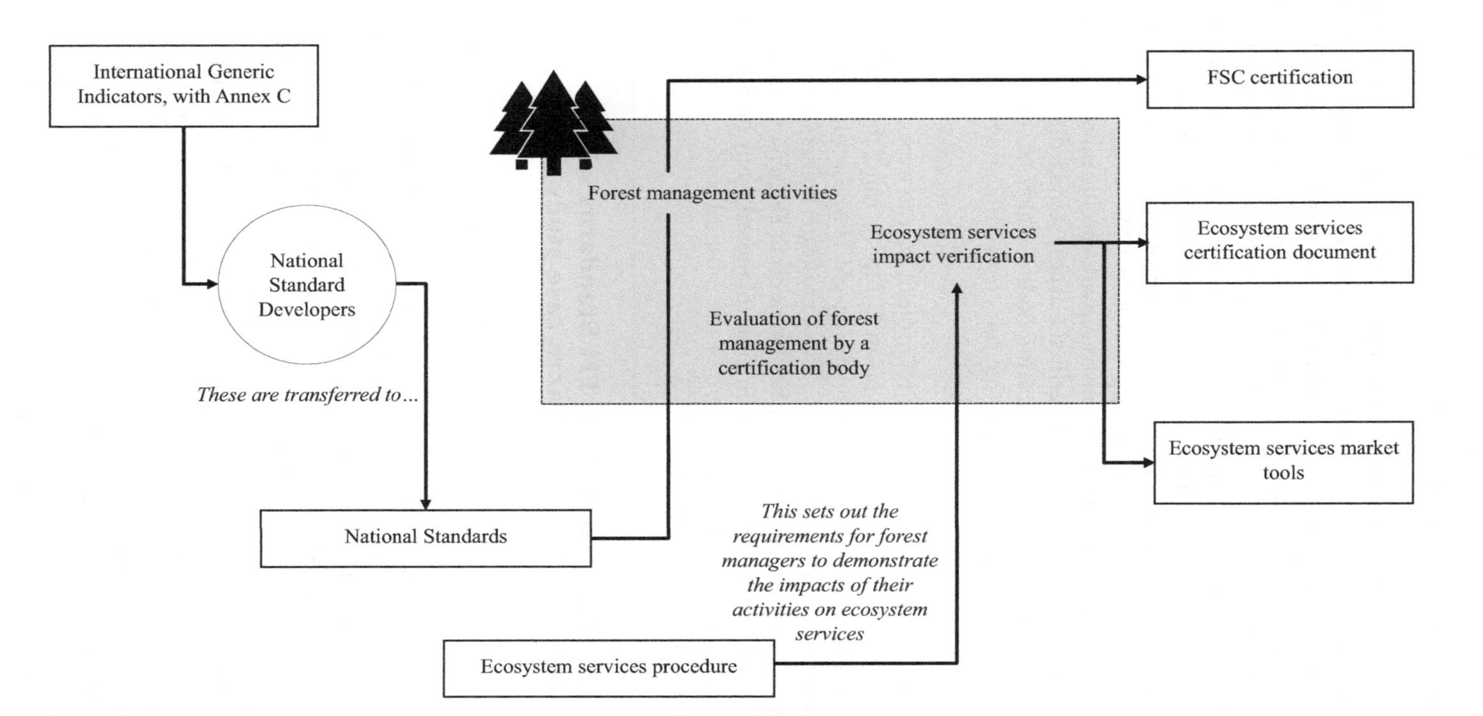


*Figure 12.1* Incorporation of ecosystem services tools in the FSC Forest Management Certification system

Source: Adapted from FSC (2017)

In summary, the Ecosystem Services Procedure includes:

- Part I on Impact Demonstration, a section that FSC Forest Management Certificate holders have to use to demonstrate in seven steps the positive impacts generated by FSC certification (these steps also include the theory of change, the selection of specific outcome indicators as well as measurements requirements for indicators and methods to be applied);
- Part II on Market Tools and the use of FSC ecosystem service claims, which clarifies the promotion of FSC-certified forests with verified FSC ecosystem service claims, the labelling and promotion of FSC-certified products, the financial sponsorship of verified FSC ecosystem service impacts and the use of external registries (with specific mention of both Gold Standard Foundation and Verified Carbon Standard);
- Part III on requirements for certification bodies, which sets out the main evaluation, surveillance and reporting rules for certification bodies willing to verify FSC ecosystem services impacts.

After the testing phase under the ForCES project, we can expect many FSC FM Certificate holders to declare their positive impacts using the Ecosystem Services Procedure and eventually integrating FSC FM Certification with existing forest carbon standards. FSC FM Certificate holders can benefit financially from ecosystem services either by merging FSC FM Certification with additional certification (like VCS, Gold Standard, etc.) or only with FSC FM Standards as a guarantee. Two of the most advanced case studies are presented in Box 12.2 and Box 12.3.

### Box 12.2 Integrating FSC FM Standards with ecosystem services: Maderacre case study

Maderacre SAC (Maderera Río Acre)[2] is a private company that has been operating in Peru since 2002. Today, Maderacre manages 220,000 hectares of forest concessions certified by the Forest Stewardship Council (FSC) for the production, processing and marketing of Certified Wood. Of the total concession, 22.7% (50,000 ha) is included in the Madre de Dios Amazon REDD Project. The South Inter-Oceanic Highway (IOH) opened in 2009, linking Brazil with the Peruvian ports, and transformed Madre de Dios from an isolated region to an agriculture production and exporting area. Before the completion of the road, to get to Cuzco required three days. Today the same road takes 14 hours. This facilitates the immigration of new settlers, mainly from rural Andean areas, and the establishment of economic activities not compatible with the forest, such as slash-and-burn agriculture, cattle ranching, etc. The project aims to reduce land pressure in the project area and its buffer zone, to ensure sustainable forest management of timber concessions (e.g., generating higher aggregated timber products), and to improve the living conditions of local communities through implementation of an avoided deforestation project to help generate more resources for the management of the area. In this sense, it is necessary to consolidate the sustainable management of the area, as is the case of concessions for harvesting of timber and non-timber resources, private conservation areas and protected natural areas.

The project started on January 1, 2009, when the validation process was completed under the Climate, Community and Biodiversity (CCB) Project Design Standards, thus ensuring the social and environmental sustainability of the proposal. This is reinforced by obtaining the Gold Level of the standard. As a complement, the project was also validated under the Verified Carbon Standard (VCS) on September 20, 2012, subsequently making the verification of 2009–2012 on 21 May 2013. Since 2012, the company sells an annual average amount of USD 125,000, covering 62.5% of the annual costs of the forest's management planning amounting to €200,000 per year. The sale of carbon credits is operated by Greenox, which retains 30% of the generated carbon credits. As of 2016, about 20% of the carbon credits generated have been sold. In the first crediting period from 2009–2012, a total of about 5 million tCO2eq was reduced. Ninety-eight percent of the carbon credits are sold in Europe. The overall satisfaction with the income generated by the sales of carbon credits is high among project shareholders.

### Box 12.3 Integrating FSC FM Standards with ecosystem services: AFP case study

AFP is the Venice Lowland Forests Association (Associazione Forestale di Pianura);[3] since 2002 it has aggregated 14 small-scale forest owners (both public and private) for a total forest area of 380 hectares in the area surrounding Venice (Italy). All AFP forests are recreational and conservation forests that serve more than 2.8 million tourists per year who spend their holidays on the beaches near Venice. Since 2016, the forests are FSC certified and, in 2017, AFP tested the first draft of the FSC Ecosystem Services Procedures to demonstrate the verified positive impacts on biodiversity, carbon and recreational services. AFP generates these services thanks to activities like reaction of new forests (tree planting), maintenance and creation of new tourist paths as well as maintenance of existing forests along with fighting invasive species, thinning activities and firefighting systems. Thanks to FSC certification, AFP increased the marketing instruments that facilitate the selling of ecosystem services. To date, AFP has sold more than €500,000 of FSC-verified ecosystem services to companies like local retailers (Supermercati Alì) as well as to international companies like IKEA and EON.

## 12.3 Certification initiatives integrating biodiversity conservation

Biodiversity conservation is automatically accounted for by Forest Management Certification as it is an inherent ecosystem service of forests.[4] Consequently, forest biodiversity services are directly or indirectly incorporated by different certification schemes for timber and carbon sequestration. Three major approaches can be distinguished in the inclusion of biodiversity conservation in certification schemes:

1 inclusion of biodiversity in voluntary forest certification (e.g., Forest Stewardship Council);
2 inclusion of biodiversity conservation as a co-benefit in forest carbon standards (e.g., Climate Community and Biodiversity Standards, the Gold Standard, etc.);
3 inclusion of biodiversity in corporate social responsibility and organizational management strategies through biodiversity offset standards (e.g., Business and Biodiversity Offsets Partnership (BBOP),[5] European Biodiversity Standard (EBS),[6] Biodiversity Benchmark certification and LIFE certification[7]).

### *12.3.1 Biodiversity co-benefits in forest carbon standards*

Biodiversity conservation is often seen as a co-benefit of forest carbon projects. For this reason, most of the forest carbon standards either acknowledge or include biodiversity conservation as a criteria for the successful certification of projects. Forest carbon standards can either be specific to biodiversity conservation (e.g., the Climate, Community and Biodiversity (CCB) Standards, the REDD+ Social and Environmental Standards (REDD+SES), etc.), or can include biodiversity requirements (e.g., the Gold Standard, Plan Vivo, etc.). The requirements of biodiversity conservation inside forest carbon standards are etherogeneous. For example, according to the Gold Standard, landscape connectivity should be mantained and 10% of the afforestation/reforestation area shall be identified and managed to protect or enhance the biological diversity of native ecosystems. The Gold Standard asks project developers to use the High Conservation Value approach (typical of FSC certification) and to carry on a biodiversity impact assessment. The CCB dedicates an entire standard section to biodiversity. Under the CCB, a biodiversity baseline – before project intervention – should be set, followed by a biodiversity impact assessment and a biodiversity monitoring plan.

### *12.3.2 Corporate responsibility and biodiversity offsetting*

One of the main threats to biodiversity is the loss of habitats. The loss and fragmentation of habitats are mainly driven by urbanization, infrastructure development and emerging agricultural frontiers (TEEB, 2012). The development and implementation of standards aiming at biodiversity conservation is an innovative strategy to ensure adequate compensation of environmental impacts of infrastructure projects. Environmental compensation is usually driven by mandatory government requirements to grant permission to construction projects having significant residual environmental impacts.[8] Nonetheless, the application of biodiversity standards can happen on a voluntary basis as a CSR strategy (Kate et al., 2004). The correct application of the mitigation hierarchy (calculate, avoid, minimize, restore and only then, offset) during the infrastructure's planning phase enables the up-scaling of compensation projects known as biodiversity offsets.

In order to guide biodiversity offsetting towards the no-net-loss of biodiversity, BBOP has developed a set of handbooks to be used at the project planning phase.[9] Moreover, the BBOP Standard on Biodiversity Offsets certifies that compensation complies with the principles, criteria and indicators to achieve the project's conservation goals. Since 2012, the BBOP standard has become an important complementary instrument and guideline for offsetting projects looking for financial support from the International Finance Corporation and its Performance Standard 6 (IFC PS6) , and from over 70 banks that have adopted the Equator Principles to finance projects[10] (BBOP, 2012).

Other standards, such as EBS, LIFE certification and Biodiversity Benchmark, support corporations in upgrading their business operations towards responsible ecological performance.

### *12.3.3 Certification initiatives integrating watershed services*

Forests contribute a lot to several water-related ecosystem services, such as pollution reduction, retention, sediment control and flood mitigation (Núñez et al., 2006). However, the maintenance and improvement of such services often bring environmental and legal restrictions and higher production costs for land managers. Therefore, worldwide, several initiatives have emerged in order to compensate the water footprint (of products and organizations) and/or to pay farmers and forest managers for their opportunity costs and for additional services generated through the adoption of improved management practices (i.e., organic farming, reforestation, sustainable forest management). As for their complexity and their site-specific characteristics, the standardization – and thus the certification – of watershed services has progressed slowly and only in some institutional contexts (such as the USA and UK, where property and use rights are better defined) (Bennett et al., 2014). We can identify four main types of standard-based certification initiatives

*Table 12.2* Categories of certification initiatives for watershed services

| *PWS type* | *Description* | *Payment/standard mechanism* | *Example* |
|---|---|---|---|
| Public regulated water-trading initiatives | Governments set a water quality standard on the total amount of pollution flowing into a body of water or watershed | Polluters collaborate to meet the standard by trading (buying and selling) pollution credits to maximize economic benefits. A third party checks the landowners against the standard. | • Long Island Sound Nitrogen Credit<br>• Salinity trading programs in NSW, Australia |
| Public regulated wetlands banking initiatives | A bank that restores, establishes or improves wetland, stream or other aquatic resource areas for the purpose of providing compensation for unavoidable impacts to aquatic resources permitted under state or local wetland regulation | The value of a bank is defined in "compensatory mitigation credits". A bank's instrument identifies the number of credits available for sale and requires the use of ecological assessment techniques to certify that those credits provide the required ecological functions. | • Wetlands banking under US Clean Water Act |
| Private water footprint offsetting initiatives | Certification schemes that allow offsetting of the water footprint through the purchasing of "water credits" generated from water projects that enhance the quality or quantity of water flow or the provision of water ecosystem services | Certified projects generate "water credits" that are sold to private and public entities that wish to compensate their footprint. The value of the credit goes to finance specific water-related projects. | • Water Restoration Certificates (USA)<br>• WWF Water Neutral Scheme |
| Private embedded water services certification | Existing forest certification standards that are trying to certify water-related ecosystem services embedded within their Sustainable Forest Management Certification | Specific addendums are included in general standards, providing additional water-friendly management practices and measurement indicators for those forest managers that would like to claim water benefits within their certification scope. | • FSC ForCES initiative<br>• FLOCert water footprint assessment |

connected with Payments for Watershed Services that can possibly be linked with forest management:

1 public-regulated Water Quality Trading initiatives;
2 public-regulated wetlands banking initiatives;
3 private water footprint offsetting initiatives;
4 private embedded water services certification.

Table 12.2 provides a brief description and examples of the four above-mentioned categories.

## Notes

1 More information on the ForCES project at the website: https://ic.fsc.org/en/what-is-fsc/what-we-do/ecosystemservices
2 Maderacre SAC. URL: www.maderacre.com
3 Associazione Forestale di Pianura. URL: www.forestedipianura.it
4 A study of the effects of forest certification on biodiversity reviews 67 projects and shows the complexity of this relationship. Available at: www.rainforest-alliance.org.
5 BBOP is an international collaboration between companies, financial institutions, government agencies and civil society organizations developing standards and guidelines to help companies to conserve biodiversity. More information at: www.bbop.forest-trends.org.
6 European Biodiversity Standard. URL: www.europeanbiodiversitystandard.eu/
7 Life certification is applied to organization management and their voluntary biodiversity conservation actions. The scheme is fully operative in Brazil at promoting private sector engagement with biodiversity conservation on a voluntary basis. The certification is under an implementation process in Paraguay, Argentina, Chile and Colombia.
8 See: Ecosystem Marketplace report. URL: www.ecosystemmarketplace.com/reports/2011_update_sbdm
9 Documents available at Business and Biodiversity Offsets Programme. URL: www.bbop.forest-trends.org.
10 The Equator Principles have become the leading voluntary standards for identifying, assessing and managing social and environmental risks related to project finance transactions. They are based on the IFC Performance Standards and the World Bank Group Environmental, Health and Safety Guidelines.

## References

BBOP (2012). *Business and Biodiversity Offsets Programme*. Standard on Biodiversity Offsets.Forest Trends, Washington, DC.

Bennett, G., Nathaniel, C., Leonardi, A. (2014). *Gaining Depth State of Watershed Investment 2014*. Forest Trends Ecosystem Marketplace, Washington, DC.

FSC (2017). ForCES: Creating Incentives to Protect Forests by Certifying Ecosystem Services. Final Report of the UN Environment/GEF-Funded Project 'Expanding FSC Certification at Landscape Level through Incorporating Additional Ecosystem Services (ID 3951)'. Forest Stewardship Council, Bonn.

Jaung, W., Putzel, L. (2013). Forest Certification for Ecosystem Services: Analysis of Market Conditions (International Market Assessment Part II). Center for International Forestry Research, Bogor.

Kate, K., Bishop, J., Rayon, R. (2004). *Biodiversity Offsets: Views, Experiences and the Business Case*. IUCN, Gland, Switzerland and Cambrindge, UK and Insights Investments, London UK.

Millennium Ecosystem Assessment (2005). *Ecosystems and Human Well-Being: Synthesis*. Island Press, Washington, DC.

Núñez, D., Nahuelhual, L., Oyarzún, C. (2006). Forests and Water: The Value of Native Temperate Forests in Supplying Water for Human Consumption. *Ecol. Econ*. 58, 606–616. doi:10.1016/j.ecolecon.2005.08.010

TEEB (2012). *The Economics of Ecosystems and Biodiversity: TEEB for Business*. International Union for Conservation of Nature and UNEP, Geneve.

Wunder, S. (2005). Payments for Environmental Services: Some Nuts and Bolts. CIFOR Occasional Paper No. 42. Bogor, Indonesia.

Part 3

# Auditing techniques in the forest sector

Chapter 13

# FSC® forest management auditing

*Mateo Cariño Fraisse*

This chapter provides a detailed look into the processes of FSC forest management auditing. After exploring the main concepts (paragraph 13.1), the chapter will first focus on the auditor and the auditing team (paragraph 13.2), and then looks at the different tasks required when performing an audit (paragraph 13.3). The described auditing approach is detailed, from the initial contact to the reporting phase, going through all the particulars of a field visit. Then, different evaluation types are enumerated and their specificities described (paragraph 13.4). Some final considerations are provided in paragraph 13.5.

## 13.1 Auditing in the FSC forest management context

Performing an audit means setting an objective and carrying out an independent evaluation of conformance to specified and agreed requirements. Often, the value of auditing to a company is just to have somebody external looking at the operations. The issues raised by the audit team should be used as valuable information for budget allocation and improving the management of forest resources. If the audit is based on an objective and professional collection of information, it provides an added value to the forest manager by helping to make management decisions, rather than just adding costs. Auditors work with a wide variety of companies and their judgements are normally aligned with industry standards, thus, from the audit results a company can see what the common best practices in the sector are. Ensuring a high level of professionalism in auditing is essential to avoid the perception of simply running another bureaucratic check.

The following concepts are crucial parts of an audit:

- Audit criteria: the set of policies, procedures or requirements used as a reference against which audit evidence is compared. In the FSC context, these are most often in the form of standards, documented agreements containing technical specifications or other precise criteria to be used consistently as rules, guidelines or characteristics, to ensure that materials, products, processes and services are fit for their purpose. Table 13.1 is a non-comprehensive list of the main audit documents for FSC Forest Management (FM) certification.

- Audit scope: the extent and boundaries of an audit. For FSC, this can include, for example, the physical locations (e.g., the Forest Management Units (FMUs)), organizational units (e.g., regional FM offices), products covered (e.g., timber, cork, resin) or processes covered (e.g., Chain of Custody (CoC)).
- Audit client: organization or person requesting an audit.
- Auditee: organization being audited.

*Table 13.1* Audit documents for FSC FM Certification

| *Document code* | *Document name* | *Description* |
|---|---|---|
| FSC-STD-01–005 V5–1 | FSC Principles and Criteria for Forest Stewardship | The 10 FSC Principles and Criteria detailed in this standard are the basis for every FSC-accredited national/regional FM standard and CB (certification body) generic FM Standards that are used for locally adapted interim FM Standards. |
| FSC-STD-01–003 | SLIMF Eligibility Criteria | This standard specifies the criteria that are used to determine if a Forest Management Enterprise (FME) qualifies as either Small or Low Intensity Managed Forest. |
| FSC-STD-30–005 | Group Forest Certification: Requirements for Group Managers | This standard presents the basic requirements to be implemented by group entities wishing to obtain a forest management group certificate. |
| FSC-POL-20–002 | Partial Certification of Large Ownerships (2000) | This policy explains the FSC's position on partial certification of large ownerships and conformance with criterion 1.6, including the various rules currently applied. This policy is applicable for all certificates that do not include all lands managed by the FME in the scope of a valid FSC Forest Management Certificate. |
| FSC-POL-20–003 | FSC Policy on the Excision of Areas from the Scope of Certification (2004) | FSC policy that defines the circumstances when it is possible to "take out" or excise specific areas from the scope of a certificate while the remaining area can receive or retain FSC certification. This policy also details criteria to be applied by the FME to monitor and mitigate negative impacts to the certified area arising from management of the excised area. |

| *Document code* | *Document name* | *Description* |
|---|---|---|
| FSC-POL-30–001 | FSC Pesticides Policy (2005) | FSC policy on the use of highly hazardous pesticides on Forest Management Units (FMUs) included within the scope of a Forest Management Certificate. |
| FSC-STD-20–006 | Stakeholder Consultation for Forest Evaluation | This FSC standard details the requirements for stakeholder consultation the CBs have to conduct to evaluate a forest manager's conformity with the requirements of the applicable Forest Stewardship Standard. |
| FSC-STD-20–007 | Forest Management Evaluation | This FSC accreditation standard sets the requirements that the CBs shall conform to in conducting FSC forest management evaluations against the FSC P&C and certification decision-making. |
| FSC-STD-20–007a | Forest Management Evaluations Addendum | Forest certification reports: this FSC standard details the requirements for forest management evaluation reports. |
| FSC-STD-20–007b | Forest Management Evaluations Addendum | Forest certification public summary reports: stakeholder involvement and transparency of the evaluation process are paramount to the FSC system. This FSC standard details the requirements for public reporting of the results of evaluations of FSC Forest Management Certificate holders. |
| FSC-DIR-20–007 | FSC Directive on FSC Forest Management Evaluations | For ease of use and to streamline document management for stakeholders, FSC has consolidated all FM-related Advice Notes into a single document. This document contains all formal advice and standard interpretations emitted from FSC International to date related to Forest Management Certification. |

Source: Own elaboration

Note that clarity is important, as sometimes the audit may have been commissioned by a financial organization (audit client) aiming to invest in a specific FM Operation (FMO, or auditee), but controlling potential risks requires them to be FSC certified while they cover the costs of the assessment.

## 13.2 The auditor and auditing team

### *13.2.1 Auditor qualifications*

The auditor's competence is demonstrated at three different levels: auditor qualities, personal attributes and educational and training background. The first level represents the underlying auditor qualities, which includes ethical conduct, fair presentation, due professional care, an objective evidence-based approach and a third-party independent approach. The latter includes that there are no conflicts of interest, which entails (i) that no direct work relations have been established with the company within the last two years, (ii) that there are no close relatives working for the company, (iii) that there is no economic relationship to the company, such as shareholder entitlements and of course (iv) that there is no personal interest in the outcome of the certification process. This first level is crucial, as the integrity of the audit depends, to a large extent, on the integrity of the auditor.

The second level contains the personal attributes. In this group, the following characteristics are useful in the FSC FM context:

a perceptiveness, or understanding the situation quickly without concluding too fast;
b persistency, or the ability to stay focused and maintain the planned course of action despite obstacles;
c other attributes that would also help to build the profile of a good auditor as being polite, honest, friendly, diplomatic, constructive, with a systematic approach and of course an open mind.

Many technicians would tend to assume that they comply with such characteristics. Nevertheless, FM audits are carried out in difficult/demanding situations with changing conditions, time and coordination constraints, cultural and language barriers, weather and other logistics and potentially adverse situations. These issues might easily turn the unadvised auditor into someone argumentative, critical, opinionated, inconsistent, inflexible and indecisive.

The third level comprises the education, experience and other technical skills. Academic training and/or a broader professional field experience related to the audit is required (e.g., forest management, ecology, biology, sociology, anthropology and wood technology). Then, good presentation abilities (both verbal and written) are key for a successful performance, together with the technical skills needed to investigate and understand audit-related aspects and the ability to organize and plan the work systematically. Lastly, specific auditor training will be needed. This includes the formal auditor training together with evaluation of acquired knowledge and field auditing experience (i.e., initially observing audits done by other experienced auditors and then conducting audits while being observed by an experienced auditor/trainer or as part of a team).

The maintenance of such an array of skills needs to be done through a mix of practice and continuous developing knowledge (e.g., participation in training sessions, attending events/conferences, personal development, teaching, attending FSC standard working groups, etc.). A good auditor should be up-to-date with general developments in the FSC system, changes and updates in the requirements and general trends in certification. The performance of the auditor is controlled by the certification body (CB).

### 13.2.2 Auditor behaviour

During their work, auditors are confronted with several psychological challenges that can be overcome only through a well-trained consciousness. The auditor should not draw conclusions until they have enough information and should view the analysed project without prejudice. It has to be clear in the auditor's mind that relations are critical for a successful audit, and that from the different perspectives there is a common goal with the client: to have a credible system that meets the FSC requirements. Thus, it is key to be open about positive and negative findings and to be capable of providing feedback on a continuous basis. The auditor should be able to maintain a friendly and open attitude, simplifying the issue, rather than the opposite (e.g., adjusting the agenda to the working habits), speaking in a clear language that allows good interaction with people at all levels and demonstrating the ability to understand the interviewed situation. The auditor should have control of body language, both their own (being relaxed and open, without showing negative impressions) and of others, demonstrating at the same time respect for cultural differences (e.g., appropriateness of handshakes) and the human factor (e.g., eye contact, appropriate distance from the interviewee, dress code and timeliness). Interviews are preferred at the interviewee's workplace; this helps the person feel at "home" and usually more relaxed, besides, most of the required documentation will usually be in the interviewee's office.

The auditor should listen actively until the end, suspending judgement but looking for hints of intentions from the interviewee, who should be speaking most of the time and not the opposite. The auditor should ensure that if clarification is needed, he or she should ask leading or close-ended questions.

### 13.2.3 The audit team

The selection of the auditing team should be based on the audit objectives, scope and criteria. The team should respond to the required knowledge and skills in line with the objectives and ensure, besides the above requirements, that the local knowledge (including language) is covered and that team determination prevails, contributing to the ability to reach conclusions. Besides the lead auditor and the supporting auditors, the team might take on technical experts (e.g., to provide specific knowledge or skills relevant to the audit scope and criteria) and auditors in training, who act as auditors under guidance and supervision of another trained auditor for the purpose of acquiring practical auditing experience and skills. Both technical experts and auditors in training can contribute to the evaluation, but shall work together with a qualified auditor. Finally, the audit team can also incorporate observers witnessing the audit, normally on behalf of a stakeholder group, the client or another party, with the scope of guiding and assisting with practical audit arrangements (e.g., logistics, ensure safety, timing, etc.). None of the observers should interfere or influence the audit process, as the audit team member's interaction and work flow is crucial for achieving good results.

## 13.3 Performing an audit

A Forest Management Certificate provides a credible guarantee that there are no major failures in conformance with the requirements of a specific standard in any FMU within the scope of the certificate. In order to provide such a guarantee, the assessments must achieve the following tasks:

- define the scope of the evaluation: analyse and describe the forest area to be evaluated in terms of one or more distinct FMUs;
- evaluate the FME's (or group entities) management system: confirm that the management system is capable of ensuring that all the requirements of the certification standard are implemented within every FMU within the scope of the evaluation;
- verify implementation of the management system: carry out sampling of sites, documents, management records and consultation with stakeholders sufficient to verify that the management system is being implemented effectively and consistently across the whole scope of the evaluation;
- identify non-conformances: carry out sampling of sites, management records and interviews with consultation stakeholders sufficient to provide a credible guarantee that there are no major non-conformances with the performance thresholds specified in the certification standard within any FMU within the scope of the evaluation.

The FSC FM audit process can be summarized into seven steps (Figure 13.1).

### *13.3.1 Initial contact and feasibility of the evaluation*

During this phase, the CB clarifies the scope and key persons in the company and checks the level of client awareness/preparation, cooperation from the auditee, availability of resources and access to documents and information. At this initial stage the CB also discusses timing, the audit team, observers/guides and site safety rules.

### *13.3.2 Audit plan and audit team*

The CB shall document, with the level of detail appropriate to the size, type and complexity of the organization and audit, the agreed objectives, scope

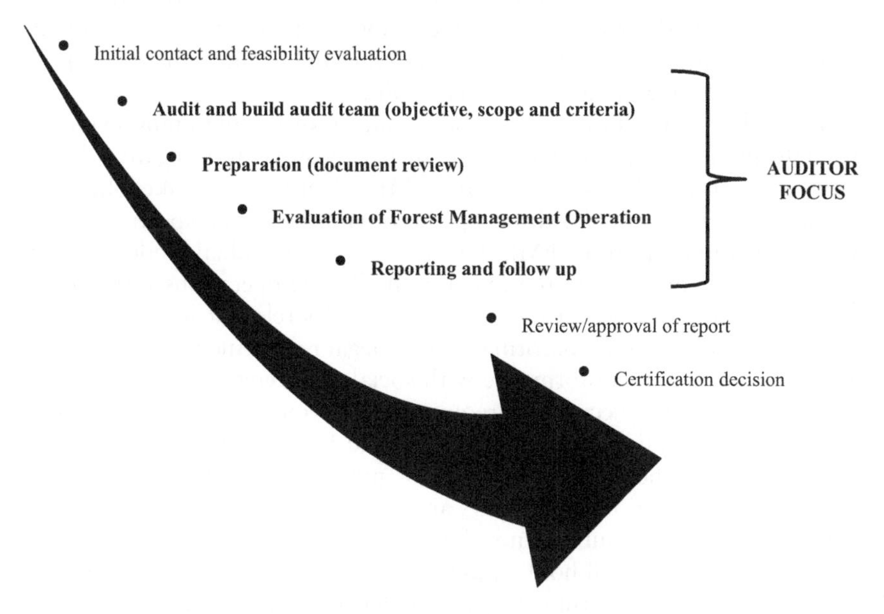


*Figure 13.1* Overview of the audit process: steps of the FSC FM audit

Source: Own elaboration

and criteria of the audit with details on the audit team, the timeline and potential staff needed (e.g., for interviews as well as other relevant information, such as logistics, audit language, audit sampling, etc.). Any objections by auditees shall be resolved before the final plan is agreed.

### *13.3.3 Preparation (document review)*

The lead auditor is responsible for the preparation phase. The auditor starts by informing the client about the audit plan (more details below) and the normative documents, and discusses the practical arrangements such as the inclusion of observers and other logistical issues. The lead auditor also needs to discuss the audit plan with the rest of the audit team, including the division of tasks and responsibilities, all the background information (e.g., about the client), normative aspects (e.g., legal requirements), working documents (e.g., standards) and logistics. The preparation of the working documents is very important for reaching the objectives in the limited time frame of the on-site visit. The working documents are the report template, client procedures and records (e.g., management plan, supply chain info, product group list) and checklists (if separated from the report template). These documents are helpful reminders and provide systematic coverage, but should not restrict the extent of the audit or its flow, which should be adapted to the practice in the company. Working documents should not make the process too formal. Major non-conformances should be communicated to the audit client (possibly for postponing or suspending the audit).

Stakeholder consultation is a fundamental tool for obtaining evidence of conformance with FSC FM requirements (positive and negative).[1] The audit team, in collaboration with the FME, identifies the stakeholders to be interviewed in order to draw up the organization's perspective, based on their relationship to the FME. Interviews with individual landowners or neighbours are more often the case for small-scale operations rather than large-scale. The stakeholder consultation provides relevant information on the conformance or non-conformance with legal requirements (Principle 1), conformance or non-conformance with social requirement (Principles 2, 3, 4 and 9), technical or economic requirements (Principles 6, 9 and 10) and environmental requirements (Principles 6 and 9). Usually, interviews with stakeholders are the primary method for public consultation. In the case of small-scale operations, interviews are between the auditor and the interviewee. In the case of public lands, large companies or controversial operations, the audit team will hold a public forum with interested stakeholders. The need for a public forum should be determined during the preparation phase, and all major stakeholders will have to receive information from the audit team before it begins. Auditors should develop a list of questions before any interview. This is especially important to maintain consistency when the same questions will be asked repeatedly with certain stakeholder groups. The line of questioning should examine what the relationship has been between the stakeholder and the FME, what concerns about specific management activities the stakeholder has had and how responsive the FME has been at resolving disputes or addressing concerns.

### 13.3.4 On-site evaluation of the FMO

#### 13.3.4.1 Initial internal team planning

Prior to meeting the FMO (Forest Management Operation, i.e., the auditee), the team will hold an internal logistic meeting(s) with the purpose of ensuring coordination, clarifying responsibilities and updating the latest information obtained with each team member. The team coordination shall be proactively maintained by the team leader throughout the audit to ensure an efficient and effective use of time, adjusting tasks when needed.

#### 13.3.4.2 Opening meeting

The initial meeting with the Forest Management Operation staff is to give them a clear understanding of what will be done during the certification assessment. It is also intended to give auditors a more specific understanding of the operation's overall management approach. In this meeting, the lead auditor presents the assessment team to the staff of the FME and outlines the steps of the field assessment and the overall certification process. After the description on the approach to forest management by the staff of the FME, the forest sites to be visited are decided on and the schedule and logistics are finalized (such as meeting places, times and personnel). Finally, the audit team gets copies of planning/monitoring documents to bring to the field and determines which ones need further review at the office.

#### *13.3.4.3 Field visit*

The audit team will complete an explicit analysis and description of the area included in the scope of the evaluation in terms of discrete FMUs and structures and systems in place for their management. The results of this analysis and description are required as the basis for subsequent evaluation of the management structure and for sampling of the population of FMUs included in the scope of the evaluation. The audit team shall complete an explicit analysis of the critical aspects of management control required to ensure that the certification standard is implemented over the full geographical area of the evaluation and the full range of management operations. In the case of applicants for group certification, the auditors shall evaluate conformance with the requirements of group certification. The evaluation should include an evaluation of the documentation and records applicable to each level of management, sufficient to confirm that management is functioning effectively and as described by the FMO. The assessment or audit must be designed and carried out to ensure that all aspects of the certification standard are audited. The goal is thus to audit the operation intensively, sampling across sites, FMUs, management situations and variables to ensure that adequate information is gathered about FMU conformance with the certification standard. As it is, in general, not possible to visit all areas, the audit team must select sites and areas to visit according to a sampling plan using both stratification and random selection. This entails weighting the most prevalent forest types and silvicultural systems and developing a cross-section of possible sites to visit. It is typically not possible to witness all stages of management happening simultaneously. To overcome this constraint, it is recommended to visit sites chronologically and those that reflect different stages of management, and choose sites where it is possible to see multiple facets of management, in order to maximize the number of management variables that can be observed per visit. A division of labour within the audit team members is recommended, together with making use of the forest manager, asking for examples for the best and worst, management challenges, etc. A specific focus has to be put on the High Conservation Value Forest (HCVF) approach taken by the FME. The audit team should ensure that the identification of attributes is correct, looking into the scope (area, all six HCVF included, etc.) and its relation to the wider area context, the impact of the operations, the data sources/quality and methodologies for collection of data, the stakeholder consultation, the mapping, the assessment of threats and risks, the management strategies to maintain and/or enhance the attributes and how monitoring is described.

While it is essential that the audit team strive to find efficiencies in response to the market competition among CBs, the level of effort allocated for FM evaluations cannot be driven by these pressures, and it is the responsibility of Accreditation Services International (ASI), as well as stakeholders, to ensure that the market is not setting the quality bar for FM evaluations. In order to uphold the integrity of the auditing function, a thorough and credible evaluation must be designed, taking into consideration the variables that influence the scope and complexity of the operation being assessed. These include the variability of organizational structure, complex tenure and/or cultural circumstances, the existence or potential for controversy, the scale,[2] risk, and intensity of forest management practices, the complexity and context of the legal and regulatory framework, the type of ownership (public, industrial, family forests, etc.), the experience in evaluating similar operations, the existence of indigenous or local community tenure rights, any ongoing tenure disputes or stakeholder issues, the availability of information required to conduct a thorough evaluation, the continuity of FME staff and operations and the expertise of the audit team related to the specific context of the evaluation. In case of group entity, the group shall be assessed in each evaluation in addition to the sampled FMUs according to the requirements specified in the group standard. Table 13.2 provides a

*Table 13.2* Non-exhaustive list of sites that could be visited to assess conformance with the FSC FM Standard

| *Seed orchards and nurseries* | *Worker accommodation and amenities* |
|---|---|
| Production forest areas in a sufficient variety of conditions (e.g., on steeper slopes, different soil conditions, different silvicultural systems) | Production forest areas: marked for thinning, recently thinned, marked for harvesting, recently harvested, one year after harvesting, five years after harvesting and ten years after harvesting |
| Areas where positive or unusual aspects of management take place (e.g., NTFP collection and animal husbandry) | Areas used by communities and/or indigenous peoples within or near the forest area |
| Sites where chemicals have been applied and stored | Water courses of different sizes, within and downstream of the forest area |
| Areas where previous non-compliances have been identified or can be equally found | Roads and forest roads of different sizes affected by forest management |
| Monitoring sites | Potential High Conservation Value Forest sites and protected areas |

Source: Own elaboration

non-exhaustive list of the sites that could be visited to assess conformance with the standard.

In order for an FME to sell Certified Wood, they should have an FSC compliant Chain of Custody (CoC) system in place. The CoC scope of a joint Forest Management/Chain of Custody (FM/CoC) certificate defines the required control systems that an FME must have in place to conform to the FSC requirements. The forest gate is the critical control point at which the CoC tracking obligation for certified products transfers from the FME to the next link in the chain, either a manufacturing facility, log broker, logging enterprise (in the case of stumpage sales), internal transfer to the manufacturing division (in the case of a vertically integrated forest products company), etc. It is required that the FME defines the forest gate(s) at the point where this control is transferred. An FME may have numerous forest gates depending on the species and products they produce.

While the analysis of the FME's performance is initiated during daily internal meetings, once field visits are completed the team must meet to establish the FME's conformance at the level of each indicator. Evaluating conformance is the primary objective of an audit. Auditors seek positive evidence to document conformance, however, any non-conformances identified shall be registered in a systematic way and presented to the client for correction. Auditors must make every effort to verify that the audit evidence obtained during the audit is true, accurate and objective. In order to verify the evidence, auditors should check for confirmation of the evidence from various sources. For example, if the documentation suggests non-conformance, the audit team will talk to the relevant staff and observe the activities in practice to obtain evidence from different sources. The auditors should also give feedback to the auditee about evidence that suggests non-conformance to ensure that the evidence has been correctly understood and interpreted. Evidence and findings should also be shared with other team members to calibrate findings and, if needed, ensure that additional evidence is gathered by other team members. As a rule, the evidence should be collected from different sources and also from different levels in the organization. Before the closing meeting, it is vital that the whole audit team has agreed on the conclusions and outcomes of the audit and that all relevant evidence has been acquired from all team members and properly considered. In order to ensure this, the audit team should always confer prior to the closing meeting with adequate time to review the audit findings and any other appropriate information collected during the audit, against the audit objectives and agree on the audit conclusions, taking into account the uncertainty inherent in the audit process.

A closing meeting should be held at the end of each audit. The closing meeting should be chaired by the lead auditor. Apart from the primary FMO contact for the audit, management representative(s) should normally be present. The structure of a closing meeting depends on the type of organization and audit, however, there are issues that need to be covered during all closing meetings. The aim of a closing meeting is to present the audit findings and conclusions in such a manner that they are understood and acknowledged by the auditee, and to present and agree on the non-conformities.

### *13.3.5 Reporting and follow up*

The audit report is the most important tangible product that an assessment team will deliver. It is essential as a means to communicate the findings of the assessment team. Designated sections of the assessment report (those not marked confidential) will become the Public Summary Report, which is posted on the FSC database[3] prior to the certificate issuance. The audit team should then ensure that the public summary sections of the report will receive wide public circulation and that the clarity and quality of their writing and explanations match up to this level.

### *13.3.6 Review/approval of report*

The report goes through several checkpoints that include the CB senior staff, external peer reviewers and the certification client.

### *13.3.7 Certification decision*

Finally, based on the outcomes reflected in the report, the final decision to award the certificate, or not, is made.

## 13.4 Types of evaluations

To obtain and maintain an FSC Forest Management Certificate, there are different types of evaluations held by the auditor team. The evaluations differ on the time, the recurrence and the contents.

### *13.4.1 Pre-assessment*

This is an audit prior to a main assessment with the objective of identifying barriers to certification (major gaps), evaluating the capacity of the applicant to implement its management systems consistently and effectively as described and preparing for the main assessment. The audit design is flexible and may be adapted to fit the type of operation being evaluated (i.e., small, simple forest vs. a large controversial operation) and the specific needs of the client. The FSC pre-assessment is required to be conducted on site and in the following cases:

- plantations larger than 10,000 ha;
- all non-plantation forest types larger than 50,000 hectares, unless the whole area meets the requirements for classification as a "small (in case of group certificate) or low intensity managed forest";

- FMUs containing High Conservation Value attributes, unless the whole area meets the requirements for classification as a "small forest".

### 13.4.2 Main assessment

The main, or full, assessment is a formal evaluation of a Forest Management Operation to determine whether they meet Forest Management Certification requirements. The evaluation is limited to the defined scope (standards and FMUs) of certification as detailed by the applicant. The assessment follows a standardized process and results in a positive or negative certification decision. In the case of a positive certification decision, the CB issues a Forest Management Certificate valid for a 5-year period.

### 13.4.3 Surveillance or annual audit

Certified operations are audited at least once annually to evaluate ongoing conformance and corrective actions within the defined scope of the certificate. One or more FMU-level site visits shall be carried out annually for all certificate holders, except in the case of those managing SLIMF operations. If an audit is not conducted each year as defined, FSC accreditation standards require the termination of the certificate.

### 13.4.4 Re-assessment

The CB may re-issue a certificate that has expired, based on the re-evaluation of the certificate holder's conformity with all aspects of the applicable Forest Management Standard and additional certification requirements.

### 13.4.5 Non-conformance verification

These evaluations are undertaken to maintain the certification for an FME that had major non-conformances that have shorter deadlines and therefore need to be remedied before the annual audit. The objective of this audit is to verify that major non-conformances have been adequately addressed by the FME.

## 13.5 Final considerations

Auditing forest management in the context of FSC certification is a challenging process. Nevertheless, many positive impacts have been described in the field by FME following the FSC rules over the last 20 years of certifying forests (Karmann and Smith, 2009). An updated and in-depth impact assessment is needed, however, to ensure good monitoring and adaptation of the system to the changing conditions, level of requirements and requests of the stakeholders involved in the system. This, together with other changes in the auditing requirements (auditor training, auditor rotation, etc.), are some of the mandates that arose from the 2014 General Assembly in Seville, so improvements are anticipated and welcome to the always evolving FSC system.

## Notes

1 The stakeholder consultation procedures are detailed in Annex F of the FSC-STD-20–006 Stakeholder Consultation for Forest Evaluation.
2 The FSC has defined streamlined procedures for the assessment and auditing of SLIMF operations. The procedural requirements vary depending upon the category of SLIMF (Small, Low Intensity, or Group SLIMFs). See FSC-STD-01–003 SLIMF eligibility criteria.
3 FSC Certificate Database. URL: https://info.fsc.org

## Reference

Karmann, M., Smith, A. (2009). FSC Reflected in Scientific and Professional Literature. Literature study on the outcomes and impacts of FSC certification. FSC IC, Bonn.

Chapter 14

# FSC® Chain of Custody auditing

*Mauro Masiero*

This chapter presents auditing standards and techniques for the assessment of Chain of Custody (CoC) according to FSC standards. Considering that only an overview of the main requirements is provided, it is advisable to make reference to official normative references when approaching audit activities.After listing the key documents to be used as references for CoC auditing (paragraph 14.1), specific auditor qualification requirements are presented (paragraph 14.2). Then, requirements for CoC auditing are presented in detail, distinguished into CoC evaluation at the forest management level (paragraph 14.3) and CoC evaluation as a separate assessment (paragraph 14.4). Paragraph 14.5 summarizes the main requirements for auditing against specific requirements, i.e., against complementary CoC standards already described in Chapter 7. The chapter ends by describing requirements for CoC audit reports and certification decisions that provide specific indications on non-compliance and their classification (paragraph 14.6), as well as with additional information regarding surveillance audits (paragraph 14.7).

## 14.1 Overview of FSC standards for Chain of Custody auditing

Auditing of CoC according to FSC standards is based on the following specific FSC standards:

- FSC-STD-20–001. Besides describing general requirements for FSC-accredited certification bodies, it indicates auditor qualification requirements (Annex 2) as well as requirements for the composition of auditing teams (Annex 3). Details are provided in paragraph 14.2 below.
- FSC-STD-20–007. This is the standard for forest management evaluations and includes requirements for CoC evaluation at the forest management level. Details are provided in paragraph 14.3 below.
- FSC-STD-20–011. This is the specific standard for CoC evaluations. Details are provided in paragraph 14.4 and beyond.

Applicable FSC CoC standards – as presented in chapter 7 – shall be considered among key documents for CoC auditing because they include the relevant requirements against which an organization is to be verified.

## 14.2 Auditor qualification requirements

In order to perform auditing activities on behalf of an FSC-accredited certification body, specific qualifications are needed according to requirements defined by Annex 2 to FSC-STD-20–001 v. 4–0. CoC Lead auditor qualifications include (FSC, 2015):

- completion of an IRCA (International Register of Certificated Auditors) registered ISO management standard auditor course or an ISO 19011 course on auditing techniques;
- completion of a formal auditor training program on FSC CoC certification and auditing standards provided by an accredited FSC training provider according to FSC-PRO-20–004;
- participation as an auditor or observer on at least four CoC audits for an FSC-accredited certification body in a three-year period, of which at least one shall be as an auditor in a main evaluation or re-evaluation, and one as an auditor in a surveillance audit.

To maintain their qualifications, auditors shall attend annual training (based on changes to relevant FSC standards), perform at least three on-site audit days every year and be involved in a witness audit every three years.

A CoC auditing team shall include a qualified CoC lead auditor, at least one team member experienced in the critical characteristics of the operational processes under evaluation and at least one team member who is fluent in the language of the area in which the evaluation takes place (as an alternative translator can be contracted). It is possible for one auditor to hold these characteristics; in this case, the CoC auditing team can consist of a single person.

## 14.3 Chain of Custody auditing in the case of Forest Management

CoC auditing, in the case of Forest Management assessment, requires the tracking and tracing system defined and implemented by the forest manager who provides sufficient evidence and guarantees that all products invoiced by the Forest Management Organization originate from the assessed Forest Management Unit (FSC, 2009). This implies, first of all, an assessment of documented procedures/instructions defined by the forest manager, including checking if they are available to, and known by, relevant staff, and if they cover all relevant system requirements, such as the identification of a figure responsible for their implementation, the specification of record keeping mechanisms, including a material volume register, staff training measures and programs and procedures for the use of FSC trademarks (if any). Then, effective implementation shall be verified by interviewing relevant personnel (e.g., forest workers, log-yard operators, transporters, administration office staff, etc.) and direct observation in the field to check if operations are really carried out as required by relevant procedures/instructions. In particular, forest gate(s) shall be identified and inspected (harvesting sites, log-yards, concentration yards, etc.) in order to verify how material can be traced back to the stump in the forest. For this, material samples can be selected and checked. During surveillance audits, this can be done by sampling invoices that include FSC-certified materials (if any) and, after verifying the they correctly include all formal elements required by FSC standards (in particular certification codes and FSC claims), asking the forest manager for evidence showing the materials can be reliably traced back to the certified Forest Management Unit. CoC auditing also requires that the certificate scope is fully clarified in terms of products and processes, with particular regard to the eventual occurrence of:

- mixing of certified materials from different Forest Management Units outside of the scope of the forest management enterprise's certificate;
- mixing of certified and uncertified materials;
- further (i.e., primary and/or secondary) processing on associated facilities;
- outsourcing of any CoC procedures or processing of FSC-certified material;
- inclusion of Non-Timber Forest Products in the certificate's scope.

In the first two cases, a separate CoC is required according to requirements defined by FSC CoC standards and described in Chapter 7. Audits shall then be performed according to FSC-STD-20–011 and a separate CoC Audit Report is required. When mixing certified and uncertified materials, the FSC standard for Controlled Wood shall also be implemented (FSC-STD-40–005). When processing occurs on associated facilities, they shall be inspected against applicable FSC standards and a separate CoC Audit Report shall be prepared. However, if associate facilities only process certified material from the scope of the forest management enterprise's certificate, a separate CoC certificate is not required. Further processing does not include log cutting or de-barking units and small portable sawmills associated with the forest management enterprise can be evaluated as part of a joint FM/CoC evaluation procedure and do not imply a separate CoC certificate is to be issued.

In the case any of CoC procedures (e.g., control, tracking, invoicing, labelling, etc.) or processing (e.g., logging, transport, etc.) of FSC-certified material that is outsourced prior to transfer of ownership at the forest gate(s), the forest management enterprise shall develop and maintain an up-to-date list of involved contractors and provide them with procedures/instructions for the performing of their activities and the management of certified material. Moreover, an outsourcing agreement – including the right for an FSC-accredited certification body to inspect the contractor – shall be signed by each contractor and recorded by the forest management enterprise.

Finally, in the case of Non-Timber Forest Products, these have to be included in the scope of the joint FM/CoC certificate for them to be labelled or promoted using the FSC trademarks. If they are further processed, the same aspects presented in the previous two pages shall be taken into consideration. FSC Advice Note 05 under DIR-20–007 states that in order to issue a CoC certificate for a Non-Timber Forest Product, an FSC-accredited certification body shall contact FSC and present a description of the product, listing all non-FSC-certified contents that may be included in it, and get written permission to proceed.

When the forest manager wishes to use FSC trademarks to label and/or promote FSC-certified products then this is to be audited as well, by checking the presence and consistency of procedures for trademark use and verifying their implementation in practice. This means interviewing staff members that are in charge of using trademarks (e.g., staff members in charge of labelling products before shipping or responsible for communication stuff such as the organization website, catalogue, brochure, etc.) to check if they are aware of procedures, as well as to verify if any use has occurred, if it was authorized by an FSC-accredited certification body and if it complies with graphic and other requirements laid down by FSC standards for trademark use by certificate holders. No trademark use for the public shall occur before a valid FSC Certificate is issued.

## 14.4 Chain of Custody auditing as a separate assessment

CoC auditing is regulated by FSC-STD-20–011 that includes general principles for evaluation as well as specific requirements distinguished into those for main and surveillance evaluations. In addition to this, *ad hoc* criteria are provided for the evaluation of organizations against specific requirements, i.e., in the case of Controlled Wood, reclaimed materials, multisite certification, and contractors for outsourced activities. Finally, guidelines for the preparation of CoC Audit Reports are supplied (see 14.5 below).

The final aim of a CoC audit is to analyse and describe single or multisite/group CoC operations to confirm that there is a control system in place ensuring that all the applicable standard requirements are properly implemented. Therefore, auditing should normally consist of the following main aspects:

- analysis and description of CoC operations;
- assessment of the management system;

- sampling (of operational sites, non-certified suppliers, contractors, documents, management records) and field assessment, including interviews with relevant personnel;
- reporting.

### 14.4.1 Analysis and description of CoC operations

The analysis and description of operational sites included under the scope of the CoC evaluation represents a preliminary step that is needed for the subsequent evaluation of the management system as well as for the identification of appropriate samples. Among other issues, the scope of the CoC system shall be discussed and defined with the organization from the beginning, in terms of:

- Sites involved as CoC operations, i.e., physical sites involved in the management of certified materials/products and material inputs used for producing certified products. These include processing and storing facilities, where material might be physically present, and facilities where relevant documents are prepared, managed or stored (e.g., administration office). Sites include those belonging to the organization under assessment, Participating Sites in case of multisite/group certification and contractors' sites if outsourced activities take place.
- Product groups identified according to requirements indicated under Chapter 7 above.
- Processes and activities regarding certified products/materials or material inputs intended for certified and performed in sites identified at the first point above.
- Applicable FSC standards and other FSC normative documents against which sites, products, processes and activities are audited.

The certification body shall be aware, in advance, of the size and complexity of CoC operations (i.e., the CoC scope), because this is needed for a proper audit scheduling (audit time as number of man/days). The CoC scope shall then be confirmed at the beginning of each assessment. In the case of multisite or group certificates, the complexity and scale of the activities within the certificate scope shall be used to assess the ability of the Central Office to manage the number of Participating Sites within the scope of the certificate and determine its annual growth limits (see Chapter 7).

### 14.4.2 Assessment of the management system

The control management of the organization shall be audited during CoC assessment to guarantee that all applicable requirements are fully implemented all over the full set of operations included within the CoC scope.

Special attention shall be paid to the identification and verification of critical control points, i.e., "places or situations in the supply chain where materials from uncertified/uncontrolled sources could enter or where certified/controlled materials could leave the system" (FSC, 2016, p. 7). These may include material purchases, storing areas for raw materials and products, sales and shipments, as well as contractor's facilities. Critical control points shall be identified based on the management system in place, material flows and CoC scope and discussed with the organization to understand whether adopted solutions comply with standard requirements and provide assurance that they are managed properly. Verification of critical control points shall also be included within field evaluation at the site.

The audit team shall assess the capacity of the organization to implement the management system as described. In particular, it shall be evaluated if technical and material resources available are appropriate (e.g., segregation materials used for distinguishing different input categories or systems/technology in use to control FSC-certified production) and human resources are adequate in number and capacity (training levels, experience, external advisors, etc.). Such characteristics can be assessed both directly, i.e., by directly evaluating the effectiveness of resources, and by means of indirect observation that might allow the auditor to identify indirect causes behind a certain non-compliance. For example, repeated mistakes in defining FSC claims for outputs can be the result of both an inappropriate procedure for recording key information (therefore an improved system shall be developed) and inadequate staff training for its implementation (therefore additional specific training shall be provided).

### 14.4.3 Field assessment at operational site

Each operational site included within the scope of the evaluation conformance to all applicable certification requirements shall be checked. This includes:

- Identification and assessment of management system documentation and relevant records to confirm that management is functioning effectively and as described by the documentation, with special reference to critical control points (see 14.4.2 above). Records might include examples of in/out invoices, processing records, evidence that suppliers' certificates have been verified, etc.
- Interviews with employees and contractors at each operational site. Variety and number of interviewed people shall be appropriately selected to cover all applicable certification requirements. Interviews shall be aimed at verifying training level as well as familiarity with, and understanding of, individual responsibilities as identified by the management system.
- Physical inspection of sites and all locations where operational activities under the scope of the certificate are carried out. When allowed,[1]

certification bodies can decide to conduct desk audits instead of physical inspections.
- Purchasing and sales documentation referring to materials and products used as inputs for FSC-certified products or otherwise related to FSC certification (e.g., to assess compliance with the Policy for the Association of Organizations with FSC (FSC, 2011)).
- Based on the previous point, evidence and confirmation that inputs classified as FSC-certified or FSC Controlled Wood were supplied with the correspondent FSC claims and certification codes by suppliers holding valid and consistent FSC Certificates.[2]
- Verification of systems for controlling FSC claims (see Chapter 7), in particular:

  (i) for the percentage and credit systems: calculations of credits and/or input percentages for each product group within the scope of the certificate;
  (ii) for the transfer system: sample records of certified outputs and evidence confirming they can be traced back to consistent inputs.

- Verification of the correct use of FSC trademarks both on-product and for promotional purposes, with the same approach already described for certified forest operations in 14.3.
- Training records, including training programs, training materials, participants lists, etc.
- Review of complaints and disputes received by the organization and/or the certification body.

Where applicable (e.g., surveillance), audits shall also include a review of Corrective Action Requests (CARs) implemented by the organization, in order to check if corresponding non-compliances were effectively addressed.

For the issuing of a CoC certificate, it is not mandatory that the organization has taken physical possession of FSC eligible inputs (see Chapter 7) before the audit time – the audit team assesses the management system in place and verifies how it works. In these cases, however, the organization is requested to notify the certification body as soon as eligible inputs are available, and a second site visit shall be performed by the certification body within three months following receipt of such a notification.

## 14.5 Audit against complementary CoC standards

When conditions for the implementation of complementary CoC standards described in Chapter 7 are in place, audits shall be performed also against these standards and according to specific requirements that are reported below.

### *14.5.1 Controlled Wood*

In the case where non-certified virgin material is sourced as an input for FSC-certified products, it shall be verified that the organization has specific procedures and measures in place to avoid wood from unacceptable sources, and instead, being sourced as controlled material. FSC-STD-20–011 distinguishes requirements for assessment against FSC-STD-40–005 v.2–1 and FSC-STD-40–005 v.3–0; here, reference is only made to the latter. In short, the requirements refer to (i) stakeholder consultation and (ii) assessment of the organization's DDS (see Chapter 7). Stakeholder consultation requirements only apply for the first evaluation and subsequent re-evaluations of the organization to FSC-STD-40–005 V3–0, as well as where material is sourced from unassessed, specified or unspecified risk areas according to the applicable FSC risk assessment (FSC, 2016). Relevant stakeholders shall be identified and invited to the consultation, and in the meantime, consultation processes shall be publicly notified in order to facilitate participation by stakeholders. Access to the information by stakeholders shall be granted at least six weeks prior to the evaluation. Consultation can be performed under the form of face-to-face meetings, phone interviews, etc., depending on the case.

The assessment of the organization's DDS aims to check whether the DDS has been designed and implemented in accordance with all applicable requirements. In particular, by means of desk and field verification the certification body shall verify if information gathered through the DDS is robust enough to confirm origin of the material, allow consistent risk assessment and designation and include effective control measures where needed. As regards risk assessment verification, the certification body shall verify the correct use of applicable FSC risk assessments. In particular, it shall verify if all applicable requirements have been followed, risk designation is based on appropriate, independent and objective sources of information and the geographic scale of the assessment is adequate to the supply area. Furthermore, it shall verify the adequateness of the risk assessment regarding the mixing of material with non-eligible inputs during transport, processing and storage before the material reaches the organization. The certification body shall also check if the risk assessment is kept up-to-date, i.e., it is reviewed and revised where necessary, including management of comments and complaints received from third parties.

As regards verification of control measures, the certification body shall verify if appropriate measures are implemented (e.g., compulsory control measures defined by the applicable NRA or alternative measures defined by the organization) and check at least a sample of each type of control measure for each type of risk identified in the DDS. Verification shall also take into account results of internal and external audits performed by the organization, any comments collected from stakeholder consultation and comments/complaints received by the certification body (FSC, 2o14).

### *14.5.2 Multisite and group certification*

In the case of multisite or group certification, at each evaluation the certification body shall assess the Central Office and a sample of the Participating Sites to assess their conformity to the applicable FSC normative requirements. In particular, the capacity of the Central Office – in terms of human, technical, material and financial resources – to manage the number of Participating Sites is assessed to define the maximum annual growth rate of the multisite/group. Compliance with any other requirement that is under the Central Office's responsibility according to FSC-STD-40–003 shall be verified, including audits of Participating Sites and corresponding findings and outputs in terms of CARs. As for assessment of Participating Sites, they shall be distinguished into two groups, i.e., Normal Risk Participating Sites and High-Risk Participating Sites.[3] For each group, the certification body shall select a minimum sample $y$ rounded to the upper whole number according to:

$$y = R\sqrt{x}$$

where $x$ is the number of Participating Sites within each group, and $R$ is a Risk Index to be determined according to a specific matrix provided by FSC-STD-20–011. By the time of a surveillance evaluation or re-evaluation, new Participating Sites that the Central Office wishes to add to the scope of a multisite or group certificate shall be considered as an independent set for the determination of the sample size. The sampling rate is the same as the one already described.

As for the composition of samples, the certification body shall include randomly selected sites and make sure the overall sample is representative of the multisite or group in terms of (i) geographic distribution, (ii) activities/products, (iii) size of sites (in terms of employees, production volumes and/or annual turnover), etc. The same Participating Sites shall not be audited in consecutive audits unless this can be justified by specific reasons, e.g., for the purposes of verifying CARs.

### *14.5.3 Reclaimed materials*

For organizations that purchase reclaimed materials to be used as inputs for FSC-certified products and have a Supplier Audit Program according to FSC-STD-40–007 (see Chapter 7), the certification body shall: (i) check that verifications were properly conducted and recorded by the organization, in compliance with all applicable requirements defined by FSC-STD-40–007 and (ii) carry out annual on-site verification audits of a sample of the supplier sites. These audits can be avoided if supplier verifications under the organization's Supplier Audit Program have been performed by other FSC-accredited certification bodies.

The certification body shall select a minimum sample $y$ rounded to the upper whole number according to:

$$y = 0.8\sqrt{x}$$

where $x$ is the number of suppliers verified by the organization in the last 12 months. The certification body can select different supplier sites from those already verified by the organization.

### 14.5.4 Outsourced operations and contractors

When one or more operations included within the CoC certificate scope are outsourced and contractors are involved, the certification body shall monitor the CoC system throughout outsourcing arrangements and check that all relevant FSC normative requirements are met. In particular, in the case of non-certified contractors, audit activities shall verify that (i) there is a list of subcontractors involved in management (i.e., storing, processing, etc.) of materials/products to be used as inputs for FSC-certified products, (ii) for each subcontractor included in the list, there is an outsourcing agreement complying with requirements defined by FSC-STD-40–004 and (iii) the organization's CoC management system includes procedures and/or instructions for the outsourced processes and these are shared with the relevant contractors, in order to control the risk of mixing and substitution or false claims by the organization or the contractor. In addition to this, the certification body shall select off-site contractors to be field-audited based on a risk assessment approach. Outsourcing arrangements and relevant contractors are classified as "high risk" when one or more of the conditions listed under the left column of Table 14.1 are in place. When these conditions occur, however, the certification body can classify contractors as "low risk" if one or more risk mitigation conditions are in place (right column of Table 14.1).

When contractors are classified as "high risk" according to Table 14.1 or whenever any risk of improper additions or mixing by the contractor is identified, the certification body shall perform a physical inspection of a sample of contractors included within the CoC scope. The certification body shall select a minimum sample $y$ rounded to the upper whole number according to:

$$y = \sqrt{x}$$

where $x$ is the number of high-risk contractors included within the organization's CoC scope.[4] The certification body can select different supplier sites from those already verified by the organization. New high-risk contractors that the organization may wish to include within the certificate scope, in the period between the certification body evaluations, shall be physically inspected by the certification body. For these purposes, the certification body shall identify a sample of the new contractors to be audited according to the same sampling criteria mentioned before.

*Table 14.1* High-risk and risk mitigation conditions for contractors

| High-risk conditions for contractors: | Risk mitigation conditions: |
|---|---|
| 1. The organization outsources all or most of the manufacturing processes of a product;<br>2. Contractor grades or sorts the material (e.g., classifying wood according to its quality, size, colour, etc.);<br>3. Contractor mixes different input materials (e.g., FSC 100% and Controlled Wood);<br>4. Contractor applies the FSC label on the product;<br>5. Contractor does not physically return the FSC-certified product following outsourced processing;<br>6. Outsourcing across national borders to countries with Transparency International's Corruption Perception Index (CPI) lower than 50. | 1. The product is permanently labelled or marked in a way that the contractor cannot alter or exchange (e.g. heat brand, printed materials);<br>2. The product is palletized, or otherwise maintained as a secure unit that is not broken apart during outsourcing;<br>3. There is no risk of contamination (e.g., intentional or accidental mixing FSC-certified materials/products with non-FSC-eligible materials/products), as the contractor handles exclusively (physically and temporally) the materials from the contracting organization;<br>4. The contractor is employed for services that do not involve manufacture or transformation of certified products (e.g., warehousing, storage, distribution, logistics).<br>5. The contractor is an FSC-certified organization that includes documented procedures for outsourcing services within the scope of its certificate. |

Source: Modified from (FSC, 2016)

During audits at the contractor's facilities, the audit team shall check records of material inputs, outputs as well as transport documentation associated with material used in the processing of FSC-certified products during outsourcing.

### *14.5.5 Transaction verification*

FSC-accredited certification bodies are requested to support ASI's transaction verification activities by reporting relevant information in a timely manner. In particular, in order to facilitate the control of false claims, certification bodies shall register in the FSC database (as non-public information): (i) organizations that reported no FSC sales since the last audit, (ii) false claims and/or fraudulent activities carried out by organizations and (iii) recommendation of organizations that should be investigated by ASI, including motivations for this (e.g., complaints received by third parties, evidence of volume mismatches between the organization and trading partners, etc.).

## 14.6 Reporting and certification decision

CoC audit reports documenting audit findings and conclusions shall be prepared according to requirements defined by FSC-STD-20–011, Part 3, which indicates minimum contents for reports. In brief, contents should include (i) a cover page, with a summary of key information (e.g., certification body's name, contact details about the assessed organization, date of the evaluation, CoC certification code, etc.), (ii) information about the certification scope, (iii) a description of the scope of the evaluation (i.e., composition and qualifications of audit team's members, audit dates, normative documents used as references), (iv) findings, including a summary of non-compliances and observations raised, as well as related CARs, (v) findings with regard

*Table 14.2* CoC major and minor non-compliances and associated correction time for CARs

| *Non-compliance type* | *Definition* | *Maximum correction time frame for CARs* | *If not addressed within given time frame* |
|---|---|---|---|
| Major non-compliance | • continues over a long period of time, or<br>• repeated or systematic, or<br>• affects a wide range of the production or a large proportion of workers, or<br>• is not adequately addressed once identified, and<br>• results in a fundamental failure to achieve the objective of the relevant CoC requirement(s). | • 3 months<br>• Exceptionally 6 months.<br>The occurrence of one or more non-compliances in a main (or renovation) audit shall prevent the certification body from issuing (or re-issuing) the certificate.<br>The occurrence of five or more major non-compliances in a surveillance audit shall lead to immediate suspension of the certificate. | Immediate suspension of the certificate |
| Minor non-compliance | • temporary lapse, or<br>• unusual/non-systematic, or<br>• its impacts are limited in their temporal and organizational scale, and<br>• does not result in a fundamental failure to achieve the objective of the relevant CoC requirement(s). | • 1 year<br>• Exceptionally 2 years<br>It is up to the audit team to verify if the number and impact of a series of minor non-compliances indicates a "systematic" failure in achieving relevant CoC requirement(s). In this case multiple minor non-compliances shall qualify as a major non-compliance. | Become major non-compliances and need to be addressed within 3 months (exceptionally 6 months) |

Source: Own elaboration

to complementary CoC requirements (if relevant) and, when needed (vi) annexes that include any additional information supporting or confirming findings (e.g., copies of sampled documents, pictures, etc.).

CoC audit reports shall be shared with, and accepted (i.e., signed), by assessed organizations they refer to. Certification decisions – i.e., the decision on whether to issue a certificate or not – shall be made by each certification body's decision-making entity (e.g., certification or scientific committee) on the basis of CoC audit reports and the evaluation of the organization's conformity to all applicable requirements. All non-compliances shall be registered in the report, making reference to the FSC requirement(s) they do not comply with. All non-compliances shall be classified as minor or major and addressed within proper time frame by means of appropriate CARs (Table 14.2). In addition to non-compliances, observations may be reported as well: they correspond to situations that may evolve into non-compliances if not properly and timely addressed by the organization. Observations represent areas of the management systems that should be improved and are to be recorded in the CoC Audit Report.

## 14.7 Surveillance evaluation

In order to confirm the organization's continued conformance to all applicable certification requirements, surveillance audits shall be performed by the certification body at least once per calendar year, but not later than 15 months after the last evaluation. Shorter intervals might be defined when needed, for example, to verify corrective actions undertaken to address major non-compliances.

Exceptions may be allowed for

> operations or sites that have not produced, labelled or sold any FSC-certified material and have not sourced controlled material or sold any FSC Controlled Wood since the previous audit. [In these cases] surveillance evaluations may be waived. However, certification bodies shall not waive more than two consecutive surveillance evaluations.
>
> (FSC, 2016, p. 13)

The organization is requested to notify the certification body as soon as they want to produce, label or sell FSC-certified products or manage (i.e., source or sell) FSC Controlled Wood. During the following surveillance audit, all records back to the previous annual surveillance shall be reviewed to confirm that the CoC system has been maintained.

In general terms, additionally to what has been already indicated in 14.4.3 above, surveillance audits should include a review of (i) changes to the certificate scope, e.g., new operations/sites, processes, products, etc., (ii) changes to the CoC management system and (iii) production of FSC-certified products and related records (e.g., volume control; see Chapter 7).

## Notes

1 Physical inspection may not be required and desk audits may be conducted instead, in the case of (FSC, 2016) (i) sites that do not take physical possession of FSC-certified materials/products or FSC Controlled Wood/controlled material in their own or rented facilities, and do not label, alter, store or re-package the products (e.g., sales offices); and (ii) storage sites of finished and labelled products, and where the certification body has confirmed through an initial physical inspection that there is no risk of mixing FSC-certified products with other materials (i.e., sites that only store FSC-certified products). Certification bodies are requested to perform physical inspection of these storage sites at least once during the five-year duration of a certificate.
2 For example, the organization might conserve hard or digital printouts of the FSC database consulted before finalizing purchases, to show the validity and scope of suppliers' certificates have been assessed.
3 High-Risk Participating Sites include sites operating a Controlled Wood verification program according to FSC-STD-40–005, a Supplier Audit Program for reclaimed materials according to FSC-STD-40–007 and/or high-risk outsourcing to non-FSC-certified contractors. Normal Risk Participating Sites include sites not conducting any of the activities considered "high risk" as above (FSC, 2016).
4 FSC-certified contractors and contractors that did not provide outsourcing services to the organization since the last certification body's evaluation do not need to be evaluated (FSC, 2016).

## References

FSC (2009). Forest Management Evaluations. FSC-STD-20–007 V.3–0 EN. Forest Stewardship Council International Centre, Bonn.
FSC (2011). Policy for the Association of Organizations with FSC. FSC-POL-01–004 V2–0 EN. Forest Stewardship Council International Centre, Bonn.
FSC (2014). FSC Directive on FSC Controlled Wood. FSC-DIR-40–005 EN. Forest Stewardship Council International Centre, Bonn.
FSC (2015). General Requirements for FSC Accredited Certification Bodies. FSC-STD-20–001 V4–0 EN. Forest Stewardship Council International Centre, Bonn.
FSC (2016). Chain of Custody Evaluations. FSC-STD-20–011 V4–0 EN. Forest Stewardship Council International Centre, Bonn.

Chapter 15

# Forest carbon project cycle and auditing

*Alex Pra and Lucio Brotto*

Auditing in the context of forest-based carbon projects requires good overall knowledge of a project cycle, its phases and actors. Thus, the aim of this chapter is to provide a look into the forest carbon project cycle, with a specific emphasis on auditing stages and procedures and requirements for third-party auditors' accreditation and approval. The first paragraph (paragraph 15.1) delineates a generic project cycle, describing its main phases and steps: project idea (15.1.1) project description (15.1.2), project validation and registration (15.1.3), project implementation and monitoring (15.1.4) and project verification and carbon offsets issuance (15.1.5). The second paragraph (paragraph 15.2) provides an introductory description of the rules for third-party auditor's accreditation and approval under the main forest carbon standards, e.g., CDM, VCS, CCB, Gold Standard and Plan Vivo.

## 15.1 Project cycle and audits

The design, organization and implementation of forest-based carbon projects can be complex and involves a wide set of actors and stakeholders. Thus, it is important to understand a generic forest carbon project timeline and "who does what" in the different phases. In order to do so, this paragraph includes an overview of:

- project timeline and phases: describing the principal project phases, and for each phase the major steps and outputs in terms of documents and activities, with specific emphasis on auditing;
- project actors: describing the actors involved in each phase of a generic forest carbon project.

In particular, to describe a generic forest carbon project, the Clean Development Mechanism (CDM) project developmental process is taken as a reference. It can be complemented with general information of the Verified Carbon Standard (VCS), the Climate, Community and Biodiversity (CCB)

Standard, the Gold Standard and the Plan Vivo system. The requirements established by each standard are detailed at a general level; therefore, for a higher level of details in approaching audit activities in forest carbon projects, it is always necessary to make reference to official normative documents of the specific standards.

In general, a forest carbon project can be divided into five major phases:

- Project Idea;
- Project Design;
- Project Validation and Registration;
- Project Implementation and Monitoring;
- Project Verification and Carbon Offsets Issuance.

A phase could happen only once during the project life (e.g., Project Idea, Design, Validation), might be periodically repeated (e.g., Verification) or even last for most of the project life (e.g., Implementation and Monitoring). In addition, two other major activities, fundraising and marketing and selling, have a long-lasting importance throughout the project lifespan to ensure the start-up and long-term economic feasibility of the project, respectively. Fundraising activities are more concentrated in the first part of the project cycle where start-up costs are higher. Marketing and selling starts after the first carbon offset registration, though the marketing plan has already been designed during the project idea and project design phases.

Figure 15.1 summarizes the project timeline and the main actors involved in each phase.

The composition of actors involved in each phase could vary considerably among projects and standards. However, the main generic actors in forest carbon projects are usually:

- Project developers: they organize and coordinate the project idea and project design phases and are ultimately responsible for the completion of the certification process. It's also very common that project developers directly sell carbon offsets.
- Consultants and services providers: they provide technical and scientific assistance to project developers. Their presence becomes essential for a credible project design and to establish a reliable monitoring system.
- Landowners: are the holders of the land title. Five main types of land ownership are possible (Hamilton et al., 2010): (i) privately owned land; (ii) government owned land; (iii) corporate entity/concession-managed state production forests; (iv) land involving collective or customary rights; and (v) mixed tenure. Landowners are relevant in the establishment of a solid legal framework and could have different levels of participation in the project design, depending on their capabilities and dependency on forest income.

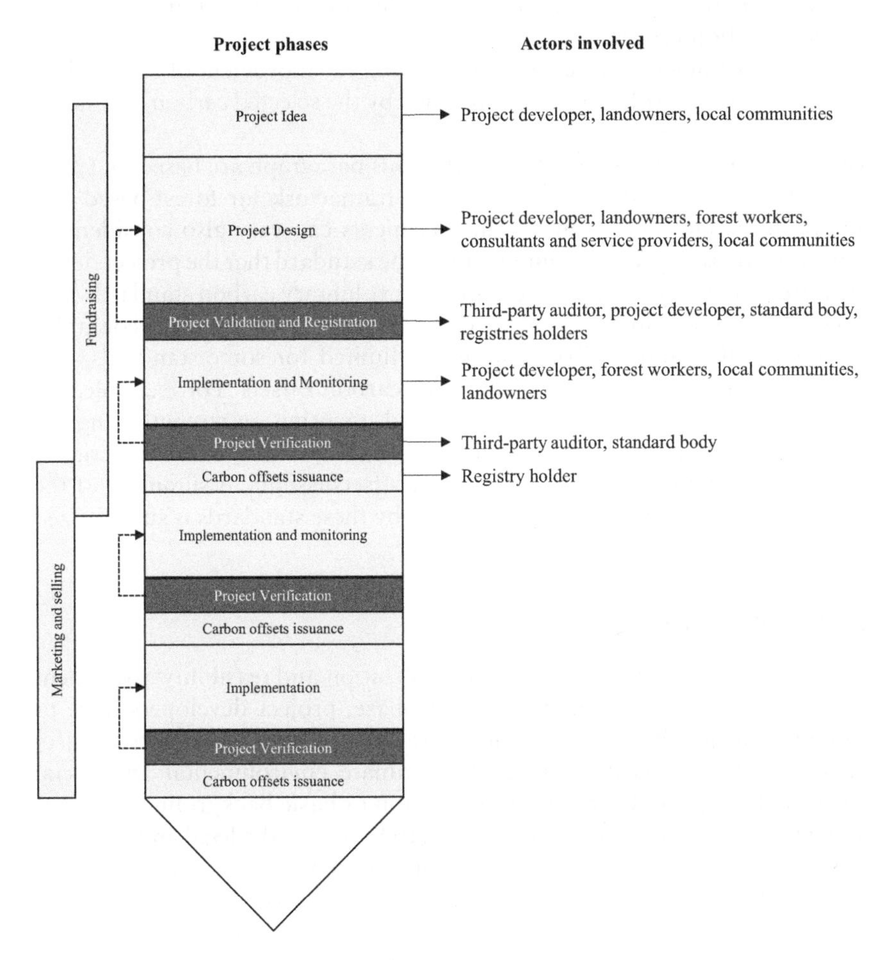


*Figure 15.1* Project phases and timeline

Source: Own elaboration

- Local communities: communities living inside or around the project area, hence directly affected by the project activities. In most of the standards, communities need to be informed and engaged in the project management from the earliest stages.
- Forestry workers: employed by the project developer or the landowners, are in charge of carrying out carbon stocks estimation, fulfillment of monitoring plan, etc.
- Third-party auditors (or validation/verification bodies): are experienced and respected auditing organizations that meet the standards requirements to conduct the validation and verification of a project.
- Standard bodies: are the organizations responsible for managing a standard program. They can be more or less involved in assuring, together

with the third-party auditors, the quality of the project and a smooth certification process.

- Registry holders: they charge for tracking the movement of sold carbon offsets. The use of registries is specified by the selected carbon standards.

The phases, steps and actors described in this paragraph are based on CDM as a reference and can serve as a generic framework for forest-based carbon projects. However, some of these elements can vary also considerably depending on the specific requirements of the standard that the project developer decides to follow, in particular, among voluntary carbon standards. It is important to remember, as explained in Chapter 8, that also the geographical applicability and eligible projects are limited for some standards, and not all standards lead to the issuance of carbon offsets. For example, the CCB Standard is a project design standard, focusing on strengthening the project design aspect to assure additional social and environmental criteria and benefits, but does not lead to carbon offsets issuing. A summary of the main project cycle and audit requirements by these standards is summarized in Table 15.1.

### 15.1.1 Project idea

The project idea comprises the conceptualization and feasibility assessment of the project. During the project idea phase, project developers have to scope the project by defining objectives, identifying suitable sites, project activities and standards to use, plus estimate environmental and social impacts of the project through the collection of basic background information. During this phase it is also essential to look into the legal and financial feasibility of the project (e.g., investment requirements, administrative and legal hurdles, development and operational costs, project risks, etc.), build partnerships and begin stakeholders' consultation and the fundraising process. The project idea phase is also an open space for project modification, where project developers can assess different intervention scenarios.

At this stage, the aim is to do a comprehensive feasibility study that will result, after a period of between six months and two years, in a document called the Project Idea Note (PIN). Although most of the voluntary carbon market standards do not ask for it, the production of the PIN is useful to sum up information gathered, to share among the project's actors and provide funders with a credible summary of the ongoing activities. PINs do not have to follow any particular format; however, a commonly used PIN format is the one developed by the World Bank's BioCarbon Fund (Box 15.1). Sometimes, the PIN is accompanied by a separate and more developed financial feasibility study, although in most of the cases it is included in the PIN.

Among the main forest carbon standards, only Plan Vivo requires the project developer to submit to the Plan Vivo Foundation a PIN (or Concept

*Table 15.1* Summary of main project cycle and audit requirements by carbon standards

| *Standard* | *PIN* | *PDD* | *Validation* | *Host country formal approval* | *Formally require stakeholders consultation* | *Registration* | *Monitoring report* | *Verification* | *Credit issuance* | *Verification and issuance cycle renewal* |
|---|---|---|---|---|---|---|---|---|---|---|
| CDM | No, but may be required in some countries by the DSA for issuing the Letter of Endorsement | Yes | Yes | Yes, Letter of Endorsement issued by DSA | Yes | CDM Registry | Yes | Yes, but verification auditor must be different than the validation auditor (with some exceptions) | Yes, CERs | Every 5 years |
| VCS | No | Yes (called Project Document) | Yes | No | No, but encouraged through an incentive system (permanence buffer) | Markit Environmental Registry and APX | Yes | Yes | Yes, VCUs | No mandatory minimum or maximum interval |
| CCB | No | Yes | Yes | No | Yes | Internal registry | Yes | Yes, the first verification can occur at the same time of the validation and according to the "rotation" rule (Par. 15.2.3) | No | Every 5 years |

(*Continued*)

*Table 15.1 (Continued)*

| *Standard* | *PIN* | *PDD* | *Validation* | *Host country formal approval* | *Formally require stakeholders consultation* | *Registration* | *Monitoring report* | *Verification* | *Credit issuance* | *Verification and issuance cycle renewal* |
|---|---|---|---|---|---|---|---|---|---|---|
| Gold Standard | No, but requires a Preliminary Review by the GS Technical Advisory Committee | Yes (called Project Documentation) | Yes (called Design Certification) plus Design Review by the GS Technical Advisory Committee | No | Yes | Markit Environmental Registry | Yes, with annual report | Yes (called performance certification), plus performance review by the GS Technical Advisory Committee | Yes, PERs | Every 5 years |
| Plan Vivo | Yes | Yes | Yes, with previous PDD approval by the Plan Vivo Foundation | No | Yes | Markit Environmental Registry | Yes, with annual report | Yes | Yes, Plan Vivo Certificates (also *ex-ante*) | Every 5 years (but recommended every 3 years) |

Source: Own elaboration

Note) that defines the main elements of the proposed projects (Box 15.2). The assessment of the Plan Vivo PIN is carried out directly by the Plan Vivo Foundation, which assesses whether the project adheres to the Plan Vivo principles and eligibility criteria and decides if it is worthwhile to continue in the project development process.

Under the Gold Standard, although a formal PIN is not required, projects must undertake at an early stage a so-called Preliminary Review. During the Preliminary Review, the project developer must upload the key project information, draft a Project Design Document and complete a stakeholder consultation report. The Gold Standard secretariat conducts a desk review of these documents in order to decide if the project has potential to conform to the Gold Standard principles and requirements. If positively assessed, the project obtains the status of "Listed" in the Gold Standard Registry, which does not provide any guarantee of certification, but only allows the project developer to proceed with the project design phase.

## Box 15.1 BioCarbon Fund PIN template

BioCarbon Fund PIN Template for Land Use, Land Use Change and Forestry (LULUCF) Projects (Including Reduced Emissions from Deforestation and Degradation Activities): http://wbcarbonfinance.org/Router.cfm?Page=DocLib&CatalogID=7110

A PIN will consist of approximately 5 pages providing indicative information on:

- the type and size of the project;
- its location;
- the anticipated total amount of greenhouse gas (GHG) reduction compared to the "business-as-usual" scenario (which will be elaborated in the baseline later on at Project Design Document level);
- the suggested crediting lifetime;
- the suggested Certified Emission Reductions (CER), Emission Reduction Unit (ERU), Removal Unit (RMU) or Verified Emission Reduction (VER) price in US$/ton CO2eq reduced from Clean Development Mechanism (CDM) or Joint Implementation (JI) projects;
- the financial structuring (indicating which parties are expected to provide the project's financing);
- the project's other socio-economic or environmental effects/benefits.

**Box 15.2 PIN template examples**

- Plan Vivo Project Idea Note template: www.planvivo.org/docs/2014_PIN_Templatedocx.docx

### 15.1.2 Project design

Once the legal, financial, environmental and social feasibility of the project has been positively assessed, the project design phase begins. This is the most time strengths consuming phase for project developers and will result in the preparation of the Project Design Document (PDD). The PDD plays a central role in project development and describes the project activity detailing the information contained in the PIN. Typically, during the preparation of the PDD, external consultants and service providers are hired to provide accuracy on some technical aspects, i.e., remote sensing, financial design or community engagement. Essential elements in the PDD are:

- background and general description of project activity;
- duration of the project activity and crediting period;
- documents clarifying land and carbon offsets property rights;
- application of the baseline, carbon quantification (including leakage) and additionality methodologies;
- non-permanence risk assessment;
- monitoring strategy;
- expected environmental and socio-economic impacts;
- stakeholders' comments;
- annexes (contact details, baseline data, monitoring plan).

The production of the PDD is a common requirement under all carbon standards. However, the different standards have different approved methodologies, set different specific requirements and can also provide supplementary tools (i.e., for additionality test, permanence risk, leakage). Methodologies define the rules that the project developer must follow to establish the project baseline, determine additionality, quantify carbon benefits and monitor the parameters used to quantify them. The project developer needs to check if the standard approved methodologies fit the project type, location and activities related. If there is not one that fits the project, the project developer can also work with the standard body to revise a pre-existing methodology or develop a new one. Thus, at this stage the choice of the standard to use must be taken very carefully by the project developer. Particular attention must be taken also at the geographical applicability (e.g., CDM does not accept project in Annex B countries; Plan Vivo accepts projects only in developing countries), and project type definition (e.g., CDM set a narrow definition of afforestation and reforestation projects, which are "direct human-induced conversion of non-forested to forested land through planting, seeding and/or the human-induced promotion of natural seed sources" on lands where no forest has existed for 50 years (afforestation) or where deforestation has taken place more recently but prior to 1989 (reforestation)). All standards provide a mandatory template for PDD (Box 15.3) and in some cases also for supplementary tools.

In the Project Design phase, formal stakeholder consultation and public comments are required under the CDM as well as under the CCB, Gold Standard and Plan Vivo. In particular, the CCB has these aspects at the core of the standard. Stakeholder consultation is not formally required under VCS, although it plays an important role as part of the risk assessment and affects the volume of offsets to set aside in the risk buffer. Most of the standards provide guidance documents on how to conduct stakeholder consultation.

PDD are found under VCS under the name Project Document (PD), while the Gold Standard refers to Project Documentation, which includes also extra mandatory documentation, i.e., Stakeholder Consultation Report, Safeguarding Principles Assessment and Defined Sustainable Development Goals (SDG) Impacts Assessment. Under the Plan Vivo system, the PDD must be accompanied by the Payment for Ecosystem Services (PES) Agreement and must be reviewed by the Plan Vivo Foundation.

### Box 15.3 PDD template examples

- CDM PDD form for afforestation and reforestation project activities v10.0: https://cdm.unfccc.int/sunsetcms/storage/contents/stored-file-20170831152800471/PDD_form06v10.pdf
- VCS Project Description Template v3.0: http://database.verra.org/sites/vcs.benfredaconsulting.com/files/VCS%20Project%20Description%20Template%2C%20v3.3.doc
- CCB Project Description template v3.0: http://verra.org/wp-content/uploads/2017/06/CCB_Project_Description_Template_CCBv3.0.docx
- CCB & VCS Project Description Template, CCB v3.0, VCS v3.3: http://verra.org/wp-content/uploads/2017/06/CCB_VCS_Project_Description_Template_CCBv3.0_VCSv3.3.docx
- Gold Standard for the Global Goals Key Project Information and Project Design Document (PDD) v1.1: https://globalgoals.goldstandard.org/wp-content/uploads/2017/06/101.1-T-PDD_rev1-.docx
- Plan Vivo PDD Template: www.planvivo.org/docs/2015-PDD-Template-Updated-May-12.docx

### 15.1.3 Project validation and registration

The PDD gathers evidence of the fulfillment of the carbon standard's requirements and is the reference document for both the audit carried out by independent third-party auditors during the project validation and the implementation and monitoring of the project. Are the expected climate benefits calculated in a sound way? Is the project methodology correct? These are some of the questions that need to be addressed during the validation. In other words, what is assessed is not the actual benefit of the project, but the way in which they have been forecast, calculated and organized, thus assessing the design of the project against a standard. The third-party auditor (or validation body) is an auditing organization that meets the standard requirements to conduct the validation of a project, based on its sectorial scope and location (see Paragraph 15.2). It is the responsibility of the project developer to identify, contract and pay the third-party auditor.

The project validation process generally consists of four phases:

- A desk review of PDD, calculation files, and GIS files.
- An on-site field visit to discuss the PDD and review documents and references, interview stakeholders and visit the project area.
- Submission to project participants of the list of Clarification Requests (requests for additional information) or Correlative Action Requests (requests for project design, description or analysis adjustment); or more in general, a draft report including indications of any data gaps or failing criteria that are not met. The project proponent then has up to 6 months to address and remedy these issues.
- Submission of a final validation report by the auditor (Box 15.4).

Validation is required by CDM as well as all carbon standards addressed in this chapter, although modalities and specific requirements vary. In some cases, it is not required and sometimes it is combined with the verification phase. For example, under the CCB, the first verification can occur at the same time as the validation.

In the case of projects under CDM, the beginning of the validation process must include a 45-day period in which the PDD is publicly consultable on the UNFCCC webpage (the so-called Global Stakeholder Process), and at its end the submission of the validation report plus the PDD to the CDM Executive Board for a completeness check and technical review. The CDM also requires the approval of the project's host country authorities with a formal Letter of Approval from the Designated National Authority (DNA). The Letter of Approval needs to state the compliance of the project activity against national rules and regulations and national criteria defining sustainable development. This process varies country by country, and although it is not required by voluntary standards, a Letter of Approval can be important to reduce the risk, legal conflicts or delays in obtaining approvals during project implementation.

The Plan Vivo system requires the PDD to be firstly approved by the Plan Vivo Foundation, which through its Technical Advisory Group coordinates a peer review of the technical aspects of the document prior to the third-party validation. In the case of the Gold Standard, this process is called Design Certification. In this case, following submission of a validation report by the third-party auditor, a so-called Design Review of Project Documentation and Validation Report is carried out by the Gold Standard Technical Advisory Committee and NGO Supporter. Successful projects obtain Gold Standard's Certified Design status.

Validation occurs only once in the project lifetime and, in the case of a positive result, is followed by the project registration in the CDM Registry in the case of projects under CDM, or in an independent publicly available database (e.g., Markit Environmental Registry, APX) in the case of projects under voluntary standards. This is also the first step to avoid double counting of carbon offsets. If the validation leads to a negative result, the project developers will have to redesign the project and undergo a new validation process.

Plan Vivo is the only standard, among the ones mentioned in this chapter, that permits the issuance of *ex-ante* offsets. Thus, once registered, the project developer can already start to establish sale contracts with the purchaser of Plan Vivo Certificates. In order to do this, the Plan Vivo Foundation requires the project developer to submit annual reports with monitoring data and updates that the Foundation needs to approve.

**Box 15.4 Validation report template examples**

- CDM validation report form for CDM project activities v3.1: https://cdm.unfccc.int/sunsetcms/storage/contents/stored-file-20180111090914799/Reg_form28v3.1.pdf
- VCS validation report template: http://database.verra.org/sites/vcs.benfredaconsulting.com/files/VCS%20Validation%20Report%20Template%2C%20v3.4.doc
- CCB validation report template: http://verra.org/wp-content/uploads/2017/06/CCB_Validation_Report_Template_CCBv3.0.docx
- CCB and VCS validation report, CCB v3.0, VCS v3.4: http://verra.org/wp-content/uploads/2017/06/CCB_VCS_Validation_Report_Template_CCBv3.0_VCSv3.4.docx

### *15.1.4 Project implementation and monitoring*

Once a project responds to an internationally accepted carbon standard, the potential to deliver offsets on the market is higher and implementation of the project activities begins. Implementation must follow what has been laid out in the PDD. How the project activities have followed the original project design will be reviewed in the verification phase. In some cases, implementation of project activities starts before the validation as a way to demonstrate project commitments, in this case, it can be called "early start implementation".

In addition, the monitoring of the project status, in accordance with the parameters and procedures laid out in the validated PDD, must be carried out. The CDM allows for changes from the approved monitoring plan only if they do not lower the level of accuracy and completeness. Voluntary carbon standards are more flexible on this, but in any case, changes must be approved with specific procedures set by the standards. Monitoring is conducted by the project developer together with the project participants and remains an ongoing activity throughout the project cycle. Most of the standards provide pre-defined monitoring report templates that the project developer can follow (Box 15.5). The results of the monitoring of the project need to be reported annually or periodically depending on the standard requirements. Monitoring is a very delicate and critical step because the actual estimation of carbon benefits of the projects will depend on the quality and accuracy of data collected and documented in the monitoring plan. In the case of forest-based carbon projects, monitoring is at the basis of the forest inventory, and needs to present a good sample design, appropriate data quality management and statistical and GIS analysis.

From an auditor's point of view, the monitoring reports in many cases present four main types of problems, that are listed hereafter with potential solutions (TÜV SUD, 2013):

- Data management, i.e., long lists of monitoring parameters, parameters defined but not monitored and lack of consistency during the reporting period. Solutions to these problems can be to carry out an initial verification and to clearly define the data management procedures in the monitoring plan.
- Quality management, i.e., lack of a quality assurance and quality control systems and lack of understanding by the project team of what is required for a climate change project. Problems that can be addressed by implementing quality assurance and quality control management systems, and defining clear roles and responsibilities in the monitoring plan.
- Uncertainty, accuracy and precision problems, i.e., high variance between planned and actual carbon stock areas. To address this risk, it is recommended to increase the number or samples or to adopt more conservative estimates.
- Undelivered amounts estimated, often caused by overly optimistic growth models, occurring from issues not considered in the design phase. One typical recommendation to avoid these types of problems is to adopt a conservative approach, be precise in site classification and try to use actual country-specific or project-specific values rather than default values.

**Box 15.5 Monitoring report template examples**

- VCS monitoring report template v3.4: http://database.verra.org/sites/vcs.benfredaconsulting.com/files/VCS%20Monitoring%20Report%20Template%2C%20v3.4.doc
- CCB monitoring report template v3.0: http://verra.org/wp-content/uploads/2017/06/CCB_Monitioring_Report_Template_CCBv3.0.docx
- CCB and VCS monitoring report, CCB v3.0, VCS v3.4: http://verra.org/wp-content/uploads/2017/06/CCB_VCS_Monitoring_Report_Template_CCBv3.0-_VCSv3.4.docx
- Gold Standard monitoring report template v1: https://globalgoals.goldstandard.org/wp-content/uploads/2017/06/101.1-T-MR.doc
- Gold Standard annual report template v1.1: https://globalgoals.goldstandard.org/wp-content/uploads/2017/06/101.1-T-MR.doc

### *15.1.5 Verification*

The verification phase aims at assessing the project implementation and monitoring, verifying the actual amount of avoided emissions or enhanced carbon stock. As for the validation, verification is conducted by an independent third-party auditor (or verification body). The validation and verification audits can be undertaken by the same auditor for the majority of voluntary carbon standards. The CDM rules require the verification audit to be undertaken by a different auditor (with some exceptions). The CCB adopted a rule under which the same validation/verification body (VVB) can undertake the validation and first verification, but the subsequent verification has to be undertaken by a different VVB. Requirements for the verification auditors are the same as with validation (see Paragraph 15.2) and it is again the project developer's responsibility to identify, contract and pay the third-party auditor.

The verification normally consists of four phases:

- desk review of the monitoring report and relevant supporting documents;
- field visit to verify the status of project activities (not always required);
- issuing of a draft verification report by the auditors, highlighting any issues emerging from the audit;
- issuing of a final verification report (Box 15.6), once the project developer solved the reported issues.

The specific modalities are defined by each standard and vary among them. Verification is usually the last step in the voluntary market before the sale of carbon offsets can happen. In most cases, the auditor quantifies and approves directly the actual emission reductions achieved by the project that can be sold as carbon offsets, without the need to go through an additional certification processes with the standard body. An exception to this is the Gold Standard, where the verification phase, called performance certification, has to be followed by a performance review carried out by the Gold Standard Technical Advisory Committee and NGO Supporter. Successful performance certification and performance review lead to the achievement of the Gold Standard Certified Project status and allow for the issuance of offsets. Also, for projects under the CDM, the verification report needs to be submitted to the CDM Executive Board for the certification and issuance of CERs and registration in the CDM registry.

In the case of projects under the CCB Standard, which does not issue carbon offsets, the focus of the verification is to verify the positive social and biodiversity impacts achieved by the projects in relation to what is planned in the PDD and monitoring plan. Many projects combine CCB with the VCS standard for issuing carbon offsets and in this case, the project developed must refer also to VCS Program requirements.

The verification and carbon offset issuance renewal differ from standard to standard. A five-year renewal cycle has been adopted by the CDM, CCB, Gold Standard, while Plan Vivo recommends a three-year cycle. VCS does not set mandatory minimum or maximum intervals.

**Box 15.6 Verification report template examples**

- CDM verification and certification report form for CDM project activities v2.1: https://cdm.unfccc.int/sunsetcms/storage/contents/stored-file-20180111090514870/Iss_form13v2.1.pdf
- CDM validation report form for renewal of crediting period for CDM project activities v2.0: https://cdm.unfccc.int/sunsetcms/storage/contents/stored-file-20171031184006073/Renewal_form05_CDM-RCPV.pdf
- VCS verification report template v3.4: http://database.verra.org/sites/vcs.benfredaconsulting.com/files/VCS%20Verification%20Report%20Template%2C%20v3.4.doc
- CCB verification report template v3.0: http://verra.org/wp-content/uploads/2017/06/CCB_Verification_Report_Template_CCBv3.0.docx
- CCB and VCS verification report, CCB v3.0, VCS v3.4: http://verra.org/wp-content/uploads/2017/06/CCB_VCS_Verification_Report_Template_CCBv3.0_VCSv3.4.docx

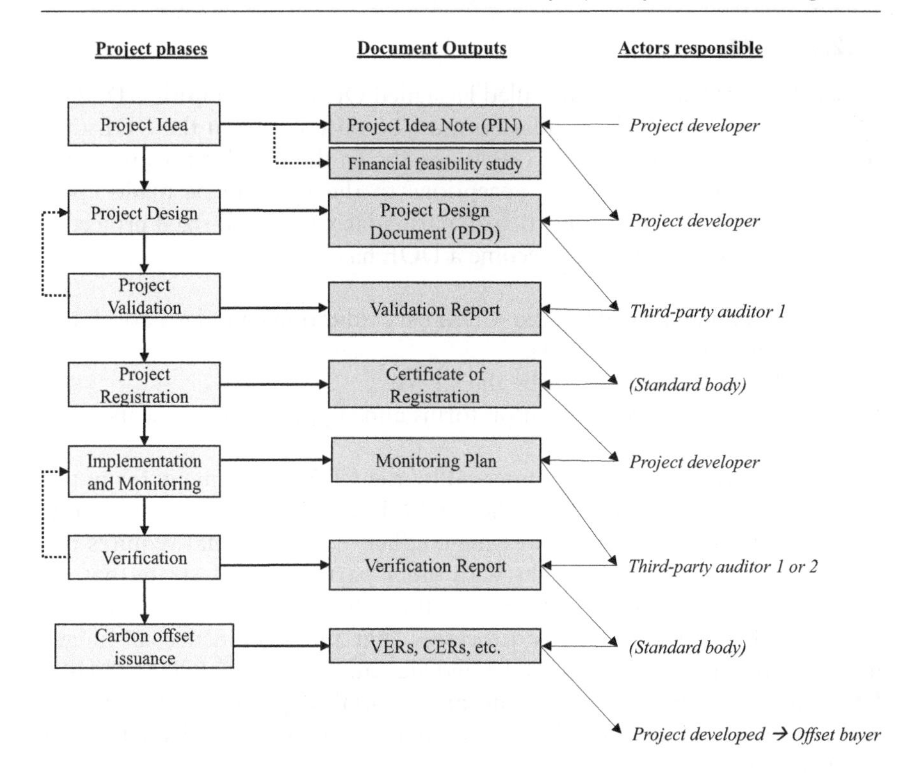


*Figure 15.2* Project phases and correspondent document outputs and responsible actors

Source: Own elaboration

Figure 15.2 summarizes the specific document outputs and responsible actors for each of the project phases described.

## 15.2 Third-party auditor's accreditation and approval

Most of the carbon standards (all of those mentioned in this chapter) require an independent third-party auditor to validate and verify the project. In the context of the carbon standard, third-party auditors are also called VVBs. A third-party auditor, or VVB, can be defined as an experienced and respected auditing organization that meets the requirements of the standards to conduct the validation or verification of a project, considering the sectorial scope and geographical location. Standards set different criteria for third-party auditor's eligibility. An overview of the criteria set for the main standards – CDM, VCS, CCB Standard, the Gold Standard and Plan Vivo – are presented in the following sections. The information presented in this paragraph is based on the standards' websites and official documents.

### 15.2.1 CDM

Under the CDM, auditors are called Designed Operational Entities (DOEs). The DOE conducting the validation must be different from the DOE that conducts the verification. This rule is specific to the CDM set to minimize conflicts of interest. However, exceptions to the rule can be made upon request to the CDM Executive Board, particularly for small-sized projects.

An organization willing to become a DOE has to:

- ensure compliance with the CDM Accreditation Standard for DOE requirements (CDM, 2014);
- pay the application fee set by the CDM;
- submit the required application forms and supporting documents.

The organization will be then assessed by the CDM Assessment Team, the CDM Accreditation Panel and the CDM Executive Board. In addition, CDM has a specific policy addressing conflict of interest, that requires the DOE to "clearly define the links with other parts to demonstrate that no conflicts of interest exist" and "demonstrate that it is not involved in any commercial, financial or other processes that might influence its judgement or endanger trust in its independence and integrity" (CDM, 2014). DOEs are accredited for one or multiple sectorial scopes (e.g., afforestation and reforestation) against which they need to demonstrate adequate team competence.

### 15.2.2 VCS

Under VCS, auditors are called VVBs and accredited to validate and verify projects as well as to assess methodology elements under the VCS methodology approval process (when there is the need to revise or develop a new methodology). Each VVB must meet key requirements set by VCS that are specific to one or multiple sectorial scopes. In order to receive VCS accreditation for validation, verification or both, VVBs have two options:

- be approved under the CDM as a DOE;
- be accredited by an International Accreditation Forum (IAF) member body such as the American National Standards Institute (ANSI) for ISO 14065.

VVBs that fill one of the two options must sign an agreement with VCS before they can perform an audit. In order to conduct an audit under the VCS standard, VVBs must hold accreditation for all sectorial scopes applicable to the audited project and methodologies (i.e., there are cases of methodologies falling under two or more sectorial scopes).

### 15.2.3 CCB Standard

Auditing organizations willing to become VVBs under the CCB have three accreditation options:

- be approved as a DOE under the CDM for the sectorial scope that matches the type of project undergoing validation or verification (e.g., "afforestation and reforestation" or "agriculture");
- be accredited as a certification body for sustainable forest management under FSC® in the geographical area of the project;
- be accredited under ISO 14065 with an accreditation scope matching the VCS AFOLU category.

To be accredited for the specific sectorial scope of the projects, CCB encourages the auditing team to include proficiency in a relevant local or regional language of the project location, relevant agriculture or forestry experience in the project country or region, as well as relevant social and ecological expertise.

As explained in Paragraph 15.1, the CCB adopts a "rotation" rule for the VVBs with respect to validation and verification of a project. The same VVB can undertake the validation and first verification (also in the case they occur at the same time), but the subsequent verification has to be undertaken by a different VVB. In addition, CCB allows the VVB to verify a project for no more than six consecutive years and can start again to verify that project only after at least three years of verification by a different VVB.

### 15.2.4 Gold Standard

The Gold Standard requires the auditors applying for GS-VVB status to meet the following requirements:

- adhere to the Gold Standard Principles;
- demonstrate no financial interests and no conflict of interest with the Gold Standard and any Gold Standard project (save for validation/verification services);
- hold an accreditation recognized by the Gold Standard; in the case of afforestation and reforestation project scopes, the organization should have the following options to be eligible:
  - be accredited under ISO 14065 for GHG activities under the ANSI-GS Accreditation Program (Scope 3);
  - be approved as a DOE status under CDM (Scope 14);
  - be accredited as an ASI-FSC certification body.

In addition, the new GS-VVB must confirm with evidence that:

- it is currently accredited under a Gold Standard Approved Accreditation;
- it has relevant competence in the scope of the certification of the audit team and individuals;
- it has the necessary capacity and resilience (e.g., resilient to changes in staff, etc.).

The GS-VVB status is granted for a 36-month period, after which a re-approval is necessary.

In order to conduct a validation or verification audit, Gold Standard requires the audit team to include at least a team leader and a lead validator/verifier, collectively demonstrating host country experience, competence in environmental and socio-economic matters, the language skills required in the country or region as well as other personal and organizational skills. In addition, in order to maintain the GS-VVB status, at least one team leader and one validator/verifier must hold the Gold Standard Training Certificate. More detailed requirements are found in the GS VVB Eligibility and Approval Requirements (Gold Standard, 2018).

### 15.2.5 Plan Vivo system

The Plan Vivo Foundation approves VVBs who:

- are experts and have appropriate experience in Payments for Ecosystem Services projects at the community level;
- have experience working in developing countries;
- have experience in verifying GHG reduction and using sustainability metrics;
- are accredited by an authority such as:
    - CDM;
    - ISO 14064;
    - FSC.

Project developers can choose one of the already accredited VVBs as well as propose a new one that should undertake a specific procedure established by Plan Vivo.

## References

CDM (2014). CDM Accreditation Standard. CDM-EB46-A02-STAN. Version 6.0. Available at: https://cdm.unfccc.int/filestorage/e/x/t/extfile-20140721152731270-accr_stan01.pdf/accr_stan01.pdf?t=VGx8cDY1OTl5fDDedGhPFzqM9YOkKpeeyCCr

Gold Standard (2018). GS-VVB Eligibility and Approval Requirements. Version 3.0. Availableat: https://globalgoals.goldstandard.org/100-gs4gg-validation-verification-body-requirements/

Hamilton, K., U. Chokkalingam and M. Bendana (2010). *State of the Forest Carbon Markets 2009: Taking Root & Branching Out*. Ecosystem Marketplace, Washington, DC.

TÜV SUD (2013). Project Cycle and Validation Experiences. Presentation at the Course "The Forest Management Auditor of Tomorrow: How to Integrate FSC® Certification and Carbon Auditing". Padova, 4–8 November 2013.

# Chapter 16

# Auditing requirements of the Sustainable Biomass Program

*Ondřej Tarabus*

This chapter provides a detailed look into the processes of the Sustainable Biomass Program (SBP) auditing requirements. Several aspects are already covered by the FSC FM Certification section (Sections 2,3, and 4 of Chapter 13) and thus will not be repeated again. After exploring the main concepts (paragraph 16.1), the chapter first focuses on the auditor and the auditing team (paragraph 16.2) and then looks at the different tasks required when performing an audit (paragraph 16.3). Several aspects of the auditing approach are very similar to FSC FM Certification and thus will not be repeated again. Then, different evaluation types are enumerated, and their specificities described (paragraph 16.4). Some final considerations are provided in paragraph 16.5.

## 16.1 Auditing in the SBP context

SBP certification is required by the main European utilities who use this system to prove the compliance with the national legislation to the regulators. This approach is used by Belgian, Danish and UK utilities, while the Dutch energy producers are still waiting for the regulator to endorse it.

The following concepts are crucial parts of an audit:

- Audit criteria: the set of policies, procedures or requirements used as a reference against which audit evidence is compared. In the SBP context, these are most often in the form of standards and instruction documents containing technical specifications. Table 16.1 is a comprehensive list of the main audit documents for SBP certification.
- Audit scope: extent and boundaries of an audit. There are two types for SBP certificate holders – trader or biomass producer. For traders, the scope includes the SBP-certified material traded, the sites where the physical address of the organization is and any possible storage areas where the material can be mixed. Biomass producer scope covers the physical locations where the biomass is produced (pellet mill, chipping sites), the

*Table 16.1* List of the main audit documents for SBP certification

| | |
|---|---|
| SBP Standard 1 | Feedstock Compliance Standard, SBP Standard 1, Version 1.0, March 2015 |
| SBP Standard 2 | Verification of SBP-Compliant Feedstock, SBP Standard 2, Version 1.0, March 2015 |
| SBP Standard 3 | Requirements for Certification Bodies, Standard 3, Version 1.0, March 2015 |
| SBP Standard 4 | Chain of Custody, SBP Standard 4, Version 1.0, March 2015 |
| SBP Standard 5 | Collection and Communication of Data, SBP Standard 5, Version 1.0, March 2015 |
| SBP Standard 6 | Energy and Carbon Balance Calculation, SBP Standard 6, Version 1.0, March 2015 |
| SBP instruction document 5A | Collection and Communication of Data |
| SBP instruction document 5B | Energy and GHG Data |
| SBP instruction document 5C | Static Biomass Profiling Data |

Source: Own elaboration

sourcing area where the feedstock can come from (country, region, FMUs) and processes covered (e.g., harvesting operations, suppliers, Chain of Custody (CoC)) and feedstock type (primary, secondary, tertiary).

- Audit client: organization or person requesting an audit.
- Auditee: organization being audited.

## 16.2 The auditor and auditing team

### 16.2.1 Auditor qualifications and auditor behaviour

Auditor qualifications and auditor behaviour are identical with FSC® FM Certification and these are already described in paragraph 13.2.

### 16.2.2 The audit team

The selection of the auditing team should be based on the audit objectives, scope and criteria. There are three basic types of audit teams; the type of team depends mostly on the type of certificate. While, for traders, the audit team consists of one auditor experienced in CoC systems, for biomass producers it might be team containing the lead auditor, local expert and carbon expert (however, in most cases the lead auditor should be able to cover the whole audit area). Additionally, the audit team can also incorporate observers witnessing the audit, normally on behalf of a stakeholder group, the client or another party, with the scope of guiding and assisting with practical audit arrangements (e.g., logistics, ensure safety, timing, etc.). None of the observers should interfere or influence the audit process, as the audit team member's interaction and work flow is crucial for achieving good results.

## 16.3 Performing an audit

An SBP certificate provides a credible guarantee that there are no major failures in conformance with the requirements of a specific standard for any material sold as SBP compliant. In order to provide such a guarantee, the assessments must achieve the following tasks:

- Define the scope of the evaluation: analyse and describe the sourcing area, define if all material comes as FSC/PEFC or SFI-certified or Supply Base Evaluation (SBE).
- Evaluate the SBP management system: confirm that the management system is capable of ensuring that all the requirements of the certification standard are implemented for all material under the scope of the evaluation.
- Verify implementation of the management system: carry out sampling of sites, documents, management records and consultation with stakeholders sufficient to verify that the management system is being implemented effectively and consistently across the whole scope of the evaluation.
- Identify non-conformances: carry out sampling of sites, management records and sufficient stakeholder consultations with interviews to provide a credible guarantee that there are no major non-conformances with the performance thresholds specified in the certification standard within any FMU within the scope of the evaluation.

### *16.3.1 Initial contact and feasibility of the evaluation, audit plan and audit team and preparation (document review)*

These topics are identical to FSC FM Certification and are already described in paragraphs 13.3.1, 13.3.2 and 13.3.3.

### *16.3.2 Stakeholder consultation*

Stakeholder consultation is done before the certification (and re-certification) audits of biomass producers; it is a fundamental tool for obtaining evidence of conformance with SBP requirements (positive and negative). The audit team should prepare a list of stakeholders for each country where the biomass is sourced. This list should contain all the registered committees or working groups developing forestry standards; the state forest service; regional NGOs that are involved or have an interest in social or environmental aspects of forest management, either at national or sub-national level, in the locality of the Supply Base (SB) to be evaluated; representatives of indigenous peoples and local communities involved or interested in forest management, either at national or sub-national level, in the locality of the SB to be evaluated; representatives of forest workers; representatives of forest harvesting industry/forest owners associations, forest research and education institutions and forest industries associations; and neighbours or local authorities.

The stakeholder consultation provides relevant information on the conformance or non-conformance with legal requirements, designation of risks for each indicator or the quality of mitigation measures implemented. Usually, e-mail notification of all stakeholders is the primary method for public consultation. In the case where there is a first SBP certificate to be issued in the country, the audit team will hold a public forum with interested stakeholders. The need for a public forum should be determined during the preparation phase, and all major stakeholders will have to receive information from the audit team before it begins.

### 16.3.3 On-site evaluation of the biomass producer

#### 16.3.3.1 Initial internal team planning, opening meeting

These topics are identical with FSC FM Certification and these are already described in paragraphs 13.3.4.1 and 13.3.4.2.

#### 16.3.3.2 Field visit

The audit team will complete an explicit analysis and description of the area included in the scope of the evaluation in terms of sourcing area, supply chains, type of feedstock sourced and structures and systems in place for their management. The results of this analysis and description are required as the basis for subsequent evaluation of the management structure and for sampling of the population of suppliers and/or FMUs included in the scope of the evaluation. The audit team shall complete an explicit analysis of the critical aspects of management control required to ensure that the certification standard is implemented over the full geographical area of the evaluation and the full range of management operations. The evaluation should include an evaluation of the documentation and records applicable to each level of management, sufficient to confirm that material is sourced from low-risk sources only. The audit must be designed and carried out to ensure that all aspects of the certification standard are audited. The goal is thus to audit the operation intensively, sampling across sites, suppliers, FMUs, management situations and variables to ensure that adequate information is gathered about biomass producer conformance with the certification standard. As it is, in general, not possible to visit all areas, the audit team must select sites and areas to visit according to a sampling plan using both stratification and random selection. This entails weighting most prevalent activity and situations and developing a cross-section of possible sites to visit.

While it is essential that the audit team strive to find efficiencies in response to the market competition among CBs, the level of effort allocated for SBP evaluations cannot be driven by these pressures, and it is the responsibility of Accreditation Services International (ASI), as well as stakeholders, to ensure that the market is not setting the quality bar for SBP evaluations. However, there are no specific requirements on how long the audit should last; it is very difficult to find common rules for this due to very different levels of effort needed in different countries. SBP evaluation should follow the sampling procedure of FSC FM defined in standard FSC-STD-20–007-V3–0.

In order for an SBP certificate holder to sell certified biomass, they shall have a Chain of Custody system implemented. Due to the fact that SBP does not have its own system in place, the certificate holder needs to obtain either FSC/PEFC or SFI CoC certification. As large amounts of SBP-certified material is produced from material already certified by FSC/PEFC or SFI, the biomass producers in the vast majority of cases already have this certification and thus it is not an obstacle. However, the traders are often required to implement two systems. A closing meeting should be held at the end of each audit and follow the same rules as FSC FM Certification.

### *16.3.4 Reporting and follow up*

The audit report is the most important tangible product that an assessment team will deliver. It is essential as a means to communicate the findings of the assessment team. The first part of the report will become the Public Summary Report, which is posted on the SBP website database[1] after the certificate issuance.

### *16.3.5 Review/approval of report*

The report goes through several checkpoints that include the CB senior staff, external peer reviewers and the certification client. First the audit report is reviewed internally and if a positive recommendation from the auditor is supported by the reviewer, the report goes to an external peer reviewer (only for certification audits with SBE), who has the final say if the certificate can be issued.

### 16.3.6 *Certification decision*

Finally, based on the outcomes reflected in the report, the final decision to award, or not, the certificate is made.

## 16.4 Types of evaluations

These topics are identical with FSC FM Certification, which was discussed in paragraph 13.4.

## 16.5 Final considerations

Auditing of the SBP system (especially when SBE is in the scope) is a challenging process. Nevertheless, in the first 5 years of its existence we can already see many positive impacts on forest management in the sourcing areas. Firstly, SBP provides an additional push for FSC or PEFC Certification of the forest (which shall have on their own a positive impact on forest management), and secondly, provides alternative solutions for subjects who can't reach FSC or PEFC Certification due to costs or administrative burden. This solution consists of SBE (risk-based system under SBP), which identifies, through risk assessment, the most critical aspects of forest management in the area and focuses on addressing these by implementing mitigation measures.

SBP is still at a very early stage of its development and will need further improvement over time, but it has already proved to be a good solution and a workable system while providing assurance that the material is sourced from responsible sources.

## Note

1 Sustainable Biomass Program accreditations and certifications. URL: https://sbp-cert.org/accreditations-and-certifications

Chapter 17

# Non-Wood Forest Products certification and auditing

*Enrico Vidale and Sabrina Tomasini*

While Chapter 10 provides a description of the standards applicable to Non-Wood Forest Products (NWFPs), this chapter provides an overview of the scopes of certification in different NWFP standards (paragraph 17.1) and clarifies two features of NWFPs that are essential for establishing meaningful auditing procedures: the ecological position (paragraph 17.2) and the habitat (paragraph 17.3). Finally, as a result of this analysis, the potential auditing indicators for the different NWFPs are presented (paragraph 17.4).

## 17.1 Comparison and coordination of certification schemes

Standards for NWFPS have different scopes, which range from sustainable forest management to organic production, from the assessment of food safety to the certification of the traditional origin of the products. All of these standards are applicable to NWFPs and they can provide numerous product benefits, such as market visibility and a premium price for producers. However, only some of them specifically target NWFPs or products collected from the wild. In particular, only the Sustainable Forest Management Standard utilizes the term "NWFP", while other standards, such as organic, prefer the term "wild". At the same time, even if the inclusion of products collected from the wild in organic certification could represent a great opportunity for NWFPs, the "wild" concept is not always communicated to the end consumer. The wild products sold in stores more frequently bear only the organic label, without specification to wild origins. This may, therefore, reveal that the organic message is often more appreciated by the consumers than the wild message.

Among the assessed standard schemes, only two, Sustainable Forest Management Standards and wild standards, include detailed ecological specifications for sustainable harvesting, while some others only indicate general principles. Table 17.1 summarizes the main scope of each certification type, whether each certification directly targets NWFPs or products collected from the wild, and the inclusion of ecological specifications.

*Table 17.1* Main scope, specificity and ecological specifications in NWFP certification

| *Certification type* | *Issue* | *Main scope* | *Specificity* | | *Inclusion of ecological specifications* |
|---|---|---|---|---|---|
| | | | *NWFPs* | *Wild* | |
| SFM (in the example of FSC) | | Assessment of sustainable forest management | Yes | – | Yes |
| Wild certification (in the example of FairWild) | | Assessment of sustainable wild harvesting | – | Yes | Yes |
| Organic (in most of the standards) | | Insurance of organic production (e.g., no use of pesticides, not contaminated areas) | – | Yes | Only general specifications |
| Environmental performance (the EU Ecolabel example) | | Assessment of low environmental impact | No | | No |
| Quality and food safety | The example of ISO | Assessment of quality of the products | No | | No |
| | The example of WHO GACP (Good Agricultural and Collection Practices) | Assurance of use of good agricultural and harvesting technical guidelines | | No (but subcategories) | Only general specifications |
| Fair Trade | | Assurance of fair prices and empowerment of producers | | No (but subcategories) | No |
| Origin, geographical indications and traditional specialties | | Assessment of the origin and the traditional know-how | | No | No |

## 17.2 The role of ecology in setting NWFP standards

NWFP standards may be implemented for a wide array of purposes, but probably the most interesting challenge to be tackled is the NWFP certification addressing sustainable harvesting. The thousands of species grouped under the term NWFP have different positions and needs within each ecosystem. Consequently, the tolerance of a given species towards harvest pressure varies enormously and the indications of sustainable wild harvest need to be based on the ecological characteristics before any certification scheme is developed and implemented in the market. Some species are inherently more resistant to harvest than others, depending on the life cycle characteristics (e.g., fruiting patterns, tolerance of shade), type of resource harvested (e.g., bark, roots, flowers, fruits), distribution and density patterns (e.g., even or clustered populations) and the population structure or size-class distribution (e.g., presence of regeneration) (Peter, 1994). Whether the harvest is sustainable or not also depends on the ecological context of the species (i.e., habitat integrity) and the local socio-economic and management context (i.e., harvest history, harvest methods, timing and frequency, spatial and time scale with short or long-term impacts) (Ticktin, 2015).

Despite the general awareness that ecological sustainability is a crucial issue in certification, it is also the most challenging to demonstrate, because information about the auto-ecology of most species as well as the impacts of harvesting is scarce. Given the great variation of ecological characteristics within the NWFP category, most certification efforts have focused on either area-based approaches, as used in timber certification, or case-by-case approaches with dedicated FSC® standards, especially for a few highly valued species, such as the Brazil nut in Bolivia, maple syrup in Canada or guadua bamboo in Colombia. However, developing case-by-case indicators is costly and time-consuming, and a group-of-species approach (e.g., for roots, leaves, fruits, etc.) has been recommended as the most sensible. An attempt has been made by the Global Non-Timber Forest Products Certification Addendum developed by NEPCon, in which ecological background information and sustainable harvest guidelines are given for groups of NWFPs based on Peters' Ecological Primer (1994). Ecological aspects and the importance of baseline information on population size and structure have also been considered in the International Standard for Sustainable Wild Collection of Medicinal and Aromatic Plants (ISSC-MAP), later incorporated in the FairWild Standard. Building on their experience, it is interesting to classify NWFPs according to their ecological position from a certification point of view. By ecological position, we mean both the type of ecosystem the NWFP grows in and whether it is dependent on a tree or not. Certification standards addressing the sustainability of the resource are based on a variety of indicators evaluating different aspects of the managed forest system. For example, forest management parameters directly affect tree-based products, while products growing in the forest ecosystem, like berries and medicinal herbs, may not be specifically addressed or monitored though basimetric area, annual growth and traditional forest indicators. An overview of the existing and the most relevant indicators for assessing sustainable harvest of the different NWFP categories may clarify which of the ecological indicators could potentially be included in future standards.

## 17.3 NWFPS and their habitats

Although the "F" in the NWFPs acronym refers to forests and other wooded lands, wild NWFPs grow in a variety of ecosystems (Figure 17.1). Firstly, despite being heavily managed, **agricultural lands** harbour a number of NWFPs, such as wild mushrooms, medicinal herbs (e.g., *Papaver rhoeas* and *Centaurea cyanus* in herbicide-free agricultural fields of temperate climates), small game (e.g., rabbits, birds) or other animals (e.g., snails, frogs, insects). These are generally regarded as undesired species, as they often interfere with controlled agricultural production, and increased intensification reduces the availability of NWFPs in this habitat. More NWFP-oriented are **agroforestry** systems. Agroforestry is a production system adopted in both tropical and temperate climates in which multipurpose trees are intercropped with understory crops to increase ecological and economic benefits (Nair, 1993); agroforestry habitats have been crucial for domestication of several NWFPs. Examples of NWFPs growing in agroforestry systems are wild mushrooms; woody climbers such as black pepper (*Piper nigrum*); understory medicinal and aromatic herbs such as American ginseng (*Panax quinquefolius*) grown under red maple trees (*Acer rubrum*); shrubs such as coffee (*Coffea arabica*), cocoa (*Theobroma cacao*) or tea (*Camellia sinensis*); medicinal and aromatic trees as canopy species, e.g., shea butter (*Vitellaria paradoxa*) or maple (*Acer* spp.); and other fruit and fodder plants (*Khaya senegalensis*). Beekeeping is also a common activity on agroforestry habitats along with the production of honey, royal jelly, propolis, wax and other bee-related products. In addition, **natural** and **semi-natural forests** of all biomes harbour NWFPs similar to agroforestry systems, such as woody climbers for food, medicine, spice or fibre; bee products; medicinal and aromatic herbs and trees; latex-, resin- and tannin-producing trees (e.g., *Larix decidua* for resin, *Schinopsis lorentzii* for tannin); game animals; wild mushrooms; and forest fruits and fodder plants. Latex-producing trees (e.g., *Hevea brasiliensis*) and palms that produce palm heart (e.g., Euterpe palms) are commonly found in **plantations**. On **pastures**, fodder plants, medicinal and aromatic herbs and bee products are harvested. **Wetlands** and **mangroves** harbour fish, crustaceans and aquatic food and fodder plants (*Phragmites australis*) or medicinal and aromatic plants such as sacred lotus (*Nelumbo nucifera*); riparian buffer zones have slippery elm (*Ulmus rubra*). Finally, NWFPs are also found in **dry lands** – prominent examples are the medicinal and aromatic incense and myrrh trees (*Boswellia sacra, Commiphora myrrha*) and the herb Devil's claw (*Harpagophytum* spp.).

The visual representation of the ecological position of NWFPs within the ecosystem in Figure 15.1 is useful for highlighting a number of aspects

*Figure 17.1* Ecological position of Non-Wood Forest Products (NWFPs) inside different environments

Note: The ellipses contain an enlargement of the picture in order to highlight the position of certain species; the dashed lines define the boundaries of the different habitats. The authors drew the picture to summarize the complexity of the ecology of NWFPs and, consequently, certification.

Source: Own elaboration

related to NWFP certification. Firstly, it clearly shows the diversity of the NWFP category compared to the wood product category. Secondly, it contributes to our understanding of how and at which level the harvest of NWFPs may impact the ecosystem from where they are extracted. Lastly, it may facilitate a clearer definition of the property and harvesting rights that are so crucial in NWFP management.

## 17.4 NWFP categories and potential indicators for their sustainable harvest

Several attempts have been made at grouping NWFPs into categories of higher or lower suitability for sustainable harvest based on their ecological impacts. It is commonly recognized that the harvesting of flowers, fruits, seeds or leaves will have fewer (immediate) negative impacts on the plant than the extraction of roots, bulbs, bark or exudates. Here, we make an attempt at linking such ecological considerations to indicators that may have an application in certification processes. Table 17.2 summarizes the socio-ecological framework for each NWFP. Table 17.3 reports the main indicators related to the NWFP type. Table 17.2 and Table 17.3 can be compared through the column "products".

In Table 17.2 we distinguish (i) tree-related products, including actual parts of the tree species (termed here "tree-based products") as well as products resulting from symbiotic relationships with trees; (ii) herb-related products, distinguishing annual/biennial from perennial plants; and (iii) ecosystem-dependent products, including animal-related products. Within each NWFP category, products are classified based on the short-term harvesting impacts on the species they originate from, i.e., causing death, damage or reduced reproduction probability. All harvest operations impact the species' population structure in the long-term – an aspect that has therefore to be monitored for all categories. The predominant use, market, habitat and ecosystem layer (tree layer, shrub layer, herb layer, forest floor, underground layer or water courses) are indicated for each category. Relevant certification indicators are selected from the existing literature and classified into forest management (FM) indicators addressing sustainability at the forest unit level, harvest and resource management (HRM) indicators addressing sustainability by suggesting appropriate harvest techniques and management interventions and ecological (E) indicators addressing sustainability at the product or species level. With no claim to be exhaustive, an example is provided to indicate possible practical applications, specifying quantitative upper and/or lower bars whenever available. For some categories, more ecological studies are available than for others. While some research has been carried out on tree-based NWFPs such as bark and resins, products derived from annual herbs seem to be particularly limited in data availability.

Certification systems may be based on the indicators proposed in Table 17.3, while keeping in mind that they are generalizations, and that

*Table 17.2* Socio-economic and ecological framework for certification of NWFPs

| *Level* | *Product type* | *Short-term harvest impact* | *Harvest method* | *Predominant habitat* | *Ecosystem layer* | *Main use and market* | *Products* |
|---|---|---|---|---|---|---|---|
| Trees | Tree-based products | Causing tree death | By removal of the whole tree | Natural forest, plantation | Tree | Beverage, leather and pharm. industry | Tannins, chemical extracts |
| | | Causing lasting damage, illness or reduced growth | By removal of or damage to structural parts | Natural forest, plantation, agroforestry | Tree | Medicinal use; food, pharmaceutical and rubber industry | Resins, gums, latex |
| | | | | | | Medicinal use; pharmaceutical and cork industry | Bark |
| | | | | | | Medicinal, fodder and ornamental use; food, pharmaceutical and ornamental industry | Branches, foliage, leaves |
| | | Causing reduced reproduction | By removal of reproductive parts | Natural forest, plantation, agroforestry | Tree | Medicinal and ornamental use; pharmaceutical and food industry | Flowers, fruits, nuts, seeds |
| | Tree-dependent products | Causing species death | By removal of the whole species | Agriculture, agroforestry, natural forest, plantation | Forest floor | Food industry | Mycorrhizal fungi |
| | | | | Natural forest, plantation, agroforestry | Tree | Medicinal use; pharmaceutical industry | Lichens |

| | | | | | | | |
|---|---|---|---|---|---|---|---|
| Herbs and palms | Annual, biennial plants | Causing plant death | By removal of the whole plant | Agriculture, agroforestry, pasture, natural forest, plantation, dry lands | Herb | Medicinal use; food and pharmaceutical industry | Whole herbs, chemical extracts |
| | | | | | Underground | Medicinal use; food, pharmaceutical and ornamental industry | Roots, bulbs, tubers |
| | | Causing reduced reproduction | By removal of reproductive and regenerating parts | Natural forest, plantation, agroforestry | Herb | Medicinal and ornamental use; food and pharmaceutical industry | Flowers, fruits, nuts, seeds |
| | | | | | | Medicinal and ornamental use; pharmaceutical and food industry | Leaves |
| | Perennial plants | Causing plant death | By removal of the whole plant | Agriculture, agroforestry, pasture, natural forest, dry lands | Herb | Medicinal use; food and pharmaceutical industry | Whole plant, chemical extracts |
| | | | | Agriculture, agroforestry, pasture, natural forest, dryland | Underground | Medicinal use; food, pharmaceutical and ornamental industry | Roots, bulbs, tubers |

(*Continued*)

*Table 17.2* (Continued)

| *Level* | *Product type* | *Short-term harvest impact* | *Harvest method* | *Predominant habitat* | *Ecosystem layer* | *Main use and market* | *Products* |
|---|---|---|---|---|---|---|---|
| | | Causing reduced reproduction | By removal of reproductive and regenerating parts | Natural forest, agroforestry | Herb | Medicinal and ornamental use; food and pharmaceutical industry | Flowers, fruits, nuts, seeds |
| | | | | Natural forest, agriculture, | Herb, shrub | Food industry | Buds, shoots |
| | | | | Natural forest, agroforestry | Herb | Medicinal and ornamental use; pharmaceutical and food industry | Leaves |
| Ecosystem | Ecosystem-dependent products | Products causing species' death | By removal of the whole individual | All | Mobile | Food and fur industry; wildlife trade | Game species |
| | | | | | | Food use and industry; wildlife trade | Insects |
| | | | | Wetlands and mangroves | Streams | Food industry | Fish and aquatic invertebrates |
| | | Impacting symbiotic relationships | By removal of products resulting from symbiotic relationships | Agroforestry, pastures, dry lands | Tree, shrub, herb | Medicinal use; food and pharmaceutical industry | Bee products |

*Table 17.3* Indicators for audit in certification of NWFPs

| *Level* | *Products* | *Relevant indicators to monitor** | *Example* | *Reference* |
|---|---|---|---|---|
| Trees | Tannins and extracts | Cut volume cannot exceed the mean annual increment (SFM) | Quebracho hardwood species | (FAO, 1998) |
| | Resins, gums, latex | Plantation established without deforestation (SFM) | Rubber (*Hevea* spp.) | (Aratrakorn et al., 2006) |
| | | Removal of external bark layers without disrupting vascular cambium (HM) | Janaguba latex (*Himanthus drasticus*) | (Baldauf et al., 2015) |
| | | Number of tapping wounds (HM) | For *Boswellia papyrifera*: max. 9 | (Eshete et al., 2012) |
| | | Minimum diameter at breast height (dbh) to start tapping (HM) | For *Boswellia papyrifera*: min. 20 cm dbh | (Eshete et al., 2012) |
| | | | Pine resin in China: min. 20–25 cm dbh | (Wang et al., 2006) |
| | | Number of tapping rounds (HM) | For *Boswellia papyrifer:* max. 7 | (Eshete et al., 2012) |
| | | Recovery time (HM) | 4 years for yearly tapped *Boswellia papyrifera* stands in Ethiopia | (Rijkers et al., 2006) |
| | | Health of the species (E) | Incidence of insect and pathogen attacks, e.g., turpentine beetles (*Dendroctomus terebrans*) on *Pinus caribaea* in Kenya | (Muga et al., 1995) |
| | | Fruit and seed production (E) | Less than 75% decline for *Boswellia papyrifera* | (Rijkers et al., 2006) |
| | | Maintain population structure over time (E) | 10 cm dbh class should be the largest class of the population (*B. papyrifera*) | (Lemenih & Kassa, 2011) |
| | Bark | Maximum bark harvest levels (HM) | 12 medicinal trees in Benin: max 50–75% of debarked trunk grants 2-year survival | (Delvaux et al., 2009) |
| | | Minimum diameter at breast height (dbh) or girth at breast height (gbh) (HM) | Cork (*Quercus suber*) in Portugal: First harvest not before 70 cm dbh | (Costa & Oliveira, 2015) |
| | | | 6 medicinal trees in India: 60 cm gbh | (Pandey, 2015) |

(*Continued*)

*Table 17.3* (Continued)

| *Level* | *Products* | *Relevant indicators to monitor** | *Example* | *Reference* |
|---|---|---|---|---|
| | | Following a specific harvest index (HM) | Cork (*Quercus suber*) in Portugal | (Costa & Oliveira, 2015) |
| | | Maximum width of longitudinal strips on main trunk (HM) | 6 medicinal trees in India: 5–8 cm wide strips | (Pandey, 2015) |
| | | Recovery time (HM) | 6 medicinal trees in India: every 18–24 months | (Pandey, 2015) |
| | | | Cork in Portugal: 9 years between harvests | (Costa & Oliveira, 2015) |
| | | Removal of external bark layers without disrupting vascular cambium (HM) | 6 medicinal trees in India | (Pandey, 2015) |
| | | Health of the species (E) | Proportion of insect and pathogen attacks | (Hall & Bawa, 1993) |
| | | Rate of wound closure (E) | 5 medicinal trees in Benin: minimum 7 cm/year | (Delvaux et al., 2010) |
| | | Maintain population structure over time (E) | | (Ticktin, 2015) |
| | Branches, foliage, leaves | Plantation established without deforestation (SFM) | Fodder shrub *Atriplex* spp. in Morocco | (Chriyaa, 2003) |
| | | Maximum level of extraction (HM) | *Khaya senegalensis* fodder | (Ticktin, 2015) |
| | | Recovery time (HM) | On *Sapindus laurifolia*: 4 years if primary branches are cut; 1 year if tertiary branches are cut | (Setty, 2015) |
| | | Yield over time (E) | | (Ticktin, 2015) |
| | | Health of the species (E) | | (Ticktin, 2015) |
| | | Number of fruit produced per inflorescence (E) | *Khaya senegalensis* foliage in Kenya | (Gaoue & Ticktin, 2008) |
| | Flowers, fruits, nuts, seeds | Leaving a proportion of fruit for seeding (HM) | 5–15% for *Sapindus laurifolia* | (Hall & Bawa, 1993; Setty, 2015) |
| | | Maximum level of extraction (HM) | For soapberries (*Sapindus laurifolia*): 86–95% | (Setty, 2015) |
| | | | For *Sclerocarya birrea* 92% | (Emanuel et al., 2005) |

| | | | | |
|---|---|---|---|---|
| | | Environmental variability stronger impacts than harvesting (E) | Soapberries (*Sapindus laurifolia*) | (Setty, 2015) |
| | | Phenological condition (E) | Proportion of crown-bearing flowers or fruit | (Hall & Bawa, 1993) |
| | | Frugivore animal populations (E) | Frugivores of *Phyllanthus emblica* in India are affected by fruit harvest | (Prasad et al., 2004) (Shahabuddin & Prasad, 2004) |
| | | Yield over time (E) | | (Ticktin, 2015) |
| | | Health of the species (E) | | (Ticktin, 2015) |
| | | Maintain population structure over time (E) | | (Ticktin, 2015) |
| | Mycorrhizal fungi | None needed | Fungi fruit body production was not adversely affected by harvest over 29 years | (Egli et al., 2006) |
| | | Monitor forest floor trampling (E) | Trampling of harvest caused less than 10 spouts and in smaller numbers | (Egli et al., 2006) |
| | Lichens | Thallus transplant (E) | For *Lobaria pulmonaria:* transplant of 5 thalli per tree | (Scheidegger et al., 2007) |
| Herbs and palms | Whole herbs, chemical extracts | Habitat protection (SFM) | | |
| | | Leaving a proportion of plants for seeding (HM) | | (Hall & Bawa, 1993) |
| | Roots, bulbs, tubers | Habitat protection (SFM) | | |
| | | Harvest only after seeding (HM) | | |
| | | Leaving a proportion of plants for seeding (HM) | | (Hall & Bawa, 1993) |
| | Flowers, fruits, nuts, seeds | Habitat protection (SFM) | For *Papaver rhoeas* flowers/seeds: set aside agricultural land | |
| | | Maximum level of extraction (HM) | For *Centaurium erythraea*: leave 20–30% of plant left on after collection | (Rodina, 2013) |

(*Continued*)

*Table 17.3* (Continued)

| *Level* | *Products* | *Relevant indicators to monitor** | *Example* | *Reference* |
|---|---|---|---|---|
| | | Leaving a proportion of plants for seeding (HM) | | (Hall & Bawa, 1993) |
| | Leaves | Maximum level of extraction (HM) | | |
| | Whole plant, chemical extracts | Habitat protection (SFM) | For *Arnica montana* in Romania: traditional hay meadow management | (Andrei & Ioan, 2009) |
| | | Leaving a proportion of herbs for seeding (HM) | | (Hall & Bawa, 1993) |
| | Roots, bulbs, tubers | Habitat protection (SFM) | Orchids require specific habitats | (Swarts & Dixon, 2009) |
| | | Protecting taproot (HM) | For Devil's claw roots (*Harpagophytum* spp.) | (Stewart & Cole, 2005) |
| | | Maximum level of extraction (HM) | For *Panax quinquefolius*: 6–8% of harvestable plants | (Ticktin et al., 2002) |
| | | Recovery time (HM) | Devil's claw roots: min. 3 years; | (Raimondo et al., 2005) |
| | | Minimum age (HM) | For *Panax quinquefolius*: min. 10 years (4 leaves) | (USDA, 2011) |
| | | Leaving a proportion of herbs for seeding (HM) | | (Hall & Bawa, 1993) |
| | | Proportion of seed vs. vegetative reproduction (E) | | (Hall & Bawa, 1993) |
| | Flowers, fruits, nuts, seeds | Maximum level of extraction (HM) | For palm fruit 70–95% | (Sampaio & Dos Santos, 2015) |
| | | | For lingonberry (*Vaccinium vitis-idaea*) and bilberry (*V. myrtillus*) fruit in Finnland: 10–15% of the biological yield | (Kangas, 1999) |
| | | | Maximum no. of flowers for *Arnica montana*: 1 flower head per plant | |
| | | Harvest only after seeding (HM) | Harvest after 20 September for golden grass (*Syngonanthus nitens*) flower stalks | (Schmidt et al., 2015) |

| | | | | |
|---|---|---|---|---|
| | | Leaving a proportion of herbs for seeding (HM) | For *Arnica montana* flowers: 50% of resource per site | (Michler, 2007) |
| | | Proportion of seed vs. vegetative reproduction (E) | | (Hall & Bawa, 1993) |
| | Buds, shoots | Timing of harvest (HM) | For palm hearts from *Archontophoenix alexandrae*: 22 months after planting | (Bovi et al., 2001) |
| | Leaves | Maximum level of extraction (HM) | *Chamaedorea* palm leaves (harvest must remain below 50% defoliation, biannually) | (Hernández-Barrios et al., 2014) |
| Ecosystem | Game species | Restricted hunting seasons (HM) | For red deer in Denmark: September to January | |
| | Insects | Habitat protection (SFM) | For monarch butterfly (*Danaus plexippus*) in Mexico: Logging restrictions | |
| | Fish and aquatic invertebrates | – | [great natural variation makes indicator development difficult] | (Aburto et al., 2015) |
| | Bee products | Amount of flowering species (E) | | (Varghese et al., 2015) |
| | | Population estimates of bee nests (E) | | (Varghese et al., 2015) |
| | | Honey yield (E) | | (Varghese et al., 2015) |

* Forest Management Indicator (SFM); Harvest and Management Indicator (HM); Ecological Indicator (E)

for each species a combination of ecological factors, harvest and management regimes has to be considered. For example, in the case of gum arabic and resin-tapping of wild trees such as *Acacia*, *Boswellia* and *Commiphora*, the risk of illness or even death of the trees is increased, so that a maximum level of ecological exploitation needs to be established. At the same time, a shift in property rights of the Acacia trees may provide incentives for sustainable harvest, i.e., from being an open-access resource, the long-term survival of which is of little interest to resin tappers with limited resources, to being the property of the landowner, who may have stronger interests in maintaining the resin-tapping at a sustainable level (Abtew et al., 2014). Another example pointing towards the need for a comprehensive assessment is cork harvesting. The legally defined levels of sustainable harvest in Portugal and Italy follow an *ad hoc* developed cork harvest index and state that the first harvest can only take place after the tree has reached a diameter at breast height of 70 cm and subsequent harvests can be done only every 9 years. It is equally important to manage and control the grazing regime in cork agroforestry systems in order to allow for natural regeneration. For some other resources, no certification indicators are needed, as is shown by Egli et al. (2006) in their long-term study on mushroom picking in Switzerland. Their results indicate that mushroom harvesting does not reduce future yields, damage the soil mycelia nor decrease species richness, irrespective of the harvesting technique (picking or cutting). Finally, auditing for NWFP certification is carried out similarly to forest certification. The main difference is related to the design and implementation of a sustainable management plan for one or more target species. The plan is generally based on one or a number of indicators that are designed to ensure the sustainable harvesting level of the target product. The cost of the design of the plan and sustainable harvesting compliance during the season are related to the complexity of the data collection for the estimation of the indicator. The use of common sustainable harvesting indicators may reduce the high certification costs that are today the biggest barrier for NWFP certification.

## 17.5 Conclusions

In this chapter, we made an attempt to define certification typologies and ecological indicators in the certification process of NWFPs. Difficulties arise due to the need to capture levels of sustainable extraction for diversified products. Our framework attempts to identify groups of NWFPs with related characteristics, which could be monitored based on similar ecological, forest management and/or harvest indicators. We also provided specific examples for each indicator and we highlighted that research on ecologically sustainable harvest levels for annual and biennial herbs need to be further explored. Certification based on ecological sustainability makes sense for many NWFPs, but not for all. For certain species, the certification of sustainable wild harvesting may be pointless, because harvesting does not seem to deplete the resources and in some cases it might even enhance the presence in the wild, as was shown in the case of wild mushrooms. Hence, a core question may be "what do we need/want to certify?" We may ask ourselves whether certification will help to achieve harvesting sustainability or if it simply represents a marketing tool.

## References

Abtew, A.A., Pretzsch, J., Secco, L., Mohamod, T.E. (2014). Contribution of Small-Scale Gum and Resin Commercialization to Local Livelihood and Rural Economic Development in the Drylands of Eastern Africa. *Forests 5*, 952–977. doi:10.3390/f5050952

Aburto, J., Cundill, G., Stotz, W. (2015). The Sustainability of Small-Scale Fishery Harvests in the Context of Highly Variable Resources, in: Shackleton, C.M., Pandey, A.K., Ticktin, T. (Eds.), *Ecological Sustainability for Non-Timber Forest Products*. Routledge, London and New York, pp. 116–125.

Andrei, S., Ioan, R. (2009). Vascular Plant Species Relationship and Grassland Productivity in Arnica Montana Habitats in the Limestone Area of Gârda De Sus Village (Apuseni Mountains – Romania) Relaţiile Interspecifice Ale Plantelor Vasculare şi Productivitatea Pajiştilor în Habit. *Cent. Eur. Agric. J. 10*, 327–332.

Aratrakorn, S., Thunhikorn, S., Donald, P.F. (2006). Changes in Bird Communities Following Conversion of Lowland Forest to Oil Palm and Rubber Plantations in Southern Thailand. *Bird Conserv. Int. 16*, 71. doi:10.1017/S0959270906000062

Baldauf, C., Corrêa, C.E., Ciampi-Guillardi, M., Sfair, J.C., Pessoa, D.D., Oliveira, R.C.F., Machado, M.F., Milfont, C.Í.D., Sunderland, T.C.H., dos Santos, F.A.M. (2015). Moving from the Ecological Sustainability to the Participatory Management of Janaguba (Himatanthus drasticus; Apocynaceae), in: Shackleton, C.M., Pandey, A.K., Ticktin, T. (Eds.), *Non-Timber Forest Products in the Global Context*. Routledge, London and New York, pp. 144–162.

Bovi, M.L.A., Saes, L.A., Uzzo, R.P., Spiering, S.H. (2001). Adequate Timing for Heart-of-Palm Harvesting in King Palm. *Hortic. Bras. 19*, 135–139. doi:10.1590/S0102-05362001000200008

Chriyaa, A. (2003). The Use of Shrubs in Livestock Feeding in Low Rainfall Areas. *L. Use, L. Cover Soil Sci. 5*.

Costa, A., Oliveira, G. (2015). Cork Oak (Quercus Suber L.): A Case of Sustainable Bark Harvesting in Southern Europe, in: Shackleton, C.M., Ticktin, T., Pandey, A.K. (Eds.), *Ecological Sustainability for Non-Timber Forest Products: Dynamics and Case Studies of Harvesting*. Routledge, London and New York, pp. 179–198.

Delvaux, C., Sinsin, B., Darchambeau, F., Van Damme, P. (2009). Recovery from Bark Harvesting of 12 Medicinal Tree Species in Benin, West Africa. *J. Appl. Ecol. 46*, 703–712. doi:10.1111/j.1365–2664.2009.01639.x

Delvaux, I., Sinsin, B., Van Damme, P. (2010). Impact of season, stem diameter and intensity of debarking on survival and bark re-growth pattern of medicinal tree species, Benin, West Africa. *Biol. Conserv. 143*(11), 2664–2671. ISSN 0006–3207, https://doi.org/10.1016 j.biocon.2010.07.009.

Egli, S., Peter, M., Buser, C., Stahel, W., Ayer, F. (2006). Mushroom Picking Does Not Impair Future Harvests: Results of a Long-Term Study in Switzerland. *Biol. Conserv. 129*, 271–276. doi:10.1016/j.biocon.2005.10.042

Emanuel, P.L., Shackleton, C.M., Baxter, J.S. (2005). Modelling the Sustainable Harvest of Sclerocarya Birrea Subsp: Caffra Fruits in the South African Lowveld. *For. Ecol. Manage. 214*, 91–103. doi:10.1016/j.foreco.2005.03.066

Eshete, A., Sterck, F.J., Bongers, F. (2012). Frankincense Production Is Determined by Tree Size and Tapping Frequency and Intensity. *For. Ecol. Manage. 274*, 136–142. doi:10.1016/j.foreco.2012.02.024

FAO (1998). Guidelines for the Management of Tropical Forests 1. The Production of Wood (FAO Forestry Paper 135). Food and Agriculture Organization of the United Nations, Rome.

Gaoue, O.G., Ticktin, T. (2008). Impacts of Bark and Foliage Harvest on Khaya senegalensis (Meliaceae) Reproductive Performance in Benin. *J. Appl. Ecol. 45*, 34–40. doi:10.1111/j.1365-2664.2007.01381.x

Hall, P., Bawa, K. (1993). Methods to Assess the Impact of Extraction of Non-Timber Tropical Forest Products on Plant Populations. *Economic Botany 47*, 234–247.

Hernández-Barrios, J.C., Anten, N.P.R., Martínez-Ramos, M. (2014). Sustainable Harvesting of Non-Timber Forest Products Based on Ecological and Economic Criteria. *J. Appl. Ecol.* n/a–n/a. doi:10.1111/1365–2664.12384

Kangas, K. (1999). Trade of Main Wild Berries in Finland. *Silva Fenn. 33*, 159–168.

Lemenih, M., Kassa, H. (2011). *Opportunities and Challenges for Sustainable Production and Marketing of Gums and Resins in Ethiopia.* Center for International Forestry Research (CIFOR), Bogor, Indonesia. ISBN: 978-602-8693-57-8

Michler, B. (2007). Conservation of Eastern European Medicinal Plants Arnica montana in Romania. Case Study Gârda de Sus Management Plan.

Muga, M.O., Kirinya, C.N., Kuria, L.G. (1995). An Assessment of Resin Tapping from Pinus Caribaea Trees Grown in Kwale District: A Report to the Chief Conservator of Forests, Forest Department.

Nair, P.K.R. (1993). An Introduction to Agroforestry. *Kluwer Academic Publishers/ ICRAF, Dordrecht.* doi:10.1016/0378–1127(95)90008-X.

Pandey, A.K. (2015). Sustainable Bark Harvesting of Important Medicinal Tree Species, India, in: Shackleton, C.M., Pandey, A.K., Ticktin, T. (Eds.), *Non-Timber Forest Products in the Global Context.* Routledge, London and New York, pp. 163–178.

Peters, C.M. (1994). *Sustainable Harvest of Non-Timber Plant Resources in Tropical Moist Forest: An Ecological Primer.* The Biodiversity Support Program/World Wildlife Fund, Washington, DC, USA.

Prasad, S., Chellam, R., Krishnaswamy, J., Goyal, S.P. (2004). Frugivory of Phyllanthus Emblica at Rajaji National Park, Northwest India. *Curr. Sci. 87*, 1188–1190.

Raimondo, D., Newton, D., Fell, C., Donaldson, J., Dickson, B. (2005). Devil's Claw Harpagophytum Spp. in South Africa: Conservation and Livelihoods Issues. *TRAFFIC Bull. 20*, 98–112.

Rijkers, T., Ogbazghi, W., Wessel, M., Bongers, F. (2006). The Effect of Tapping for Frankincense on Sexual Reproduction in Boswellia Papyrifera. *J. Appl. Ecol. 43*, 1188–1195. doi:10.1111/j.1365–2664.2006.01215.x.

Rodina, K. (Ed.) (2013). Common Useful Wild Plants in Central Europe. Promoting Traditional Collection and Use of Wild Plants to Reduce Social and Economic Disparities in Central Europe. TRAFFIC and WWF, Budapest, Hungary.

Sampaio, M.B., Dos Santos, F.A.M. (2015). Harvesting of Palm Fruits Can Be Ecologically Sustainable: A Case of Buriti (Mauritia Flexuosa; Arecaceae) in Central Brazil, in: Shackleton, C.M., Pandey, A.K., Ticktin, T. (Eds.), *Non-Timber Forest Products in the Global Context*. Routledge, London and New York, pp. 73–89.

Scheidegger, C., Stähli, I., Ellenberger, A. (2007). Nachhaltige Wildsammlung und in situ-Vermehrung der geschützten Flechtenart Lobaria pulmonaria. Heilpflanzenforsch. der Weleda – Weleda-Naturals GmbH, 13–19.

Schmidt, I.B., Figueiredo, I.B., Ticktin, T. (2015). Sustainability of Golden Grass Flower Stalk Harvesting in the Brazilian Savanna, in: Shackleton, C.M., Pandey, A.K., Ticktin, T. (Eds.), *Non-Timber Forest Products in the Global Context*. Routledge, London and New York, pp. 199–214.

Setty, R.S. (2015). The Sustainability of Soapberry (Sapindus laurifolia Vahl) Fruit Harvest of the Soliga Community in South India, in: Shackleton, C.M., Pandey, A.K., Ticktin, T. (Eds.), *Ecological Sustainability for Non-Timber Forest Products*. Routledge, London and New York, pp. 126–143.

Shahabuddin, G., Prasad, S. (2004). Assessing Ecological Sustainability of Non-Timber Forest Produce Extraction: The Indian Scenario. *Conserv. Soc. 2*, 235–250.

Stewart, K.M., Cole, D. (2005). The Commercial Harvest of Devil's Claw (Harpagophytum Spp.) in Southern Africa: The Devil's in the Details. *J. Ethnopharmacol. 100*, 225–236. doi:10.1016/j.jep.2005.07.004.

Swarts, N.D., Dixon, K.W. (2009). Terrestrial Orchid Conservation in the Age of Extinction. *Ann. Bot. 104*, 543–556. doi:10.1093/aob/mcp025.

Ticktin, T. (2015). The Ecological Sustainability of Non-Timber-Forest-Product Harvest: Principles and Methods, in: Shackleton, C.M., Pandey, A.K., Ticktin, T. (Eds.), *Ecological Sustainability for Non-Timber Forest Products*. Routledge, London and New York, pp. 31–53.

USDA (2011). *Guidance Wild Crop Harvesting*. Traffic International, Cambridge, UK.

Varghese, A., Nath, S., Leo, R., Thomas, S.G. (2015). The Road to Sustainable Harvests in Wild Honey Collection: Experiences from the Nilgiri Biosphere Reserve, Western Ghats, India, in: Shackleton, C.M., Pandey, A.K., Ticktin, T. (Eds.), *Ecological Sustainability for Non-Timber Forest Products*. Routledge, London and New York, pp. 103–116.

Wang, Z., Calderon, M.M., Carandang, M.G. (2006). Effects of Resin Tapping on Optimal Rotation Age of Pine Plantation. *J. For. Econ. 11*, 245–260. doi:10.1016/j.jfe.2005.10.001.

# Index

Note: Page numbers in italics indicate figures; page numbers in bold indicate tables.

For Product Safety Concerns and Information please contact our EU representative GPSR@taylorandfrancis.com Taylor & Francis Verlag GmbH, Kaufingerstraße 24, 80331 München, Germany

Printed and bound by CPI Group (UK) Ltd, Croydon, CR0 4YY
01/05/2025
01858391-0001